ALLE ZEIT WACH 1842

R. J. Srám V. Bulyzhenkov
L. Prilipko Y. Christen (Eds.)

Ethical Issues of Molecular Genetics in Psychiatry

Springer-Verlag
Berlin Heidelberg New York
London Paris Tokyo
Hong Kong Barcelona

Srám, Radim J., Dr.
Psychiatric Research Institute
18103 Prague 8
Czechoslovakia

Bulyzhenkov, Victor, Dr.
World Health Organization
Avenue Appia
1211 Geneva 27
Switzerland

Prilipko, Leonid, Dr.
World Health Organization
Avenue Appia
1211 Geneva 27
Switzerland

Christen, Yves, Dr.
Fondation IPSEN pour la Recherche
Thérapeutique
30, rue Cambronne
75737 Paris CEDEX
France

ISBN-13: 978-3-642-76431-8 e-ISBN-13: 978-3-642-76429-5
DOI: 10.1007/ 978-3-642-76429-5

Softcover reprint of the hardcover 1st edition 1991

Bookbinding: J. Schäffer, 6718 Grünstadt
2127/3140/543210 – Printed on acid-free paper

Preface

Over the past few years, genetics research has been in a phase of remarkably sustained and continuous revolution. The advent of "new genetics" of recombinant DNA has resulted in new discoveries occurring at a breath-taking pace, many of which have important clinical implications, for example, in new approaches to the diagnosis and treatment of hemoglobinopathies, cystic fibrosis and some forms of muscular dystrophies. Recent findings of psychiatric relevance have included the localization of the genes for Huntington's chorea and the use of DNA probes in predictive testing. Advances have been achieved in the understanding of the molecular biology of Alzheimer's disease, and at least some familiar forms of the condition appear to be linked to a gene of chromosome 21.

Taking into account current achievements in molecular genetics as well as future findings, it can be predicted that the application of new genetic technologies is likely to lead to ethical problems in practical psychiatry. In order to initiate discussions aiming to generate ideas and develop the background for future consensus in the complex area of ethics relating to the application of molecular approaches in the study of psychiatric disorders, the World Health Organization, in collaboration with the IPSEN Foundation, organized in Brno, Czechoslovakia, June 11–12, 1990, an international conference to review knowledge related to molecular genetic studies in psychiatry, with particular reference to ethical problems.

Participants at the conference represented a broad range of medical disciplines (including genetics, molecular biology, psychiatric genetics, jurisprudence and ethics) and the meeting gave an excellent opportunity for a comprehensive discussion of the problems. The discussions which took place at the conference were used to further revise the working papers which are presented here in the hope that they will help in strengthening the exchange of information in the scientific community and thereby contribute to progress in the development of ethical standards in medicine for the benefit of mankind.

Although the papers included in this volume have been produced in the framework of WHO's review of knowledge about the genetics of mental disorders, they express the views of the individual authors rather than a consensus of the participants at the conference. Also, the views do not necessarily represent the decisions or stated policy of the World Health Organization or the IPSEN Foundation.

The editors wish to express their appreciation to Mr G. Dafour, President of the IPSEN Foundation, and to Dr N. Sartorius, Director of the Division of Mental Health, World Health Organization, for the encouragement which they provided for this work. It is also a pleasure to thank Father T. Martinec, Abbot of the Augustinian Monastery in Brno, Czechoslovakia, Assistant Professor V. Orel, Director of the Mendel Museum in Brno, and the Psychiatric Research Institute in Prague for the organization of the conference, as well as to Mrs J. Mervaillie of the IPSEN Foundation for her assistance in the organization of the meeting and Mrs M.-L. Gage for her editorial assistance.

R.J. Srám · V. Bulyzhenkov · L. Prilipko · Y. Christen

Contents

Contributors

Baron, M.
Department of Medical Genetics, New York State Psychiatric Institute, 722 West 168th Street, New York, NY 10032, USA

Berg, K.
Institute of Medical Genetics, University of Oslo, Ullevål Hospital, Oslo, Norway

Crow, T. J.
Division of Psychiatry, Clinical Research Centre, Watford Road, Harrow, Middlesex, HA1 3UJ, UK

de Filippi, C.
Instituto A. Stensen, Viale don Minzoni 25A, 50129, Florence, Italy

Friedmann, Th.
Muriel Whitehill Chair of Biomedical Ethis Professor, Center for Molecular Genetics and Dept. of Pediatrics

Gage, F. H.
Department of Neurosciences, UCSD, School of Medicine, La Jolla, CA 92093, USA

Gill, M.
Institute of Psychiatry, DeCrespigny Park, Denmark Hill, London SE5 8AF, UK

Ivanov, V. I.
USSR National Research Centre of Medical Genetics, The USSR Academy of Medical Sciences, Moscow, USSR

Kimura, R.
International Bioethics Project, Kennedy Institute of Ethics, Georgetown University, Washington, DC 20057, USA, and Waseda University, Tokyo, Japan

Mattei, J-F.
Centre de Diagnostic Prénatal, Hôpital d'Enfants de la Timone, 13385 Marseille CEDEX 5, France

McGuffin, P.
Department of Psychological Medicine, University of Wales, College of Medicine, Heath Park, Cardiff, Wales

Paces, V.
Institute of Molecular Genetics, Czecholsovak Academy of Sciences, 16637 Prague 6, Czechoslovakia

Sargeant, M.
Department of Psychological Medicine, University of Wales, College of Medicine, Heath Park, Cardiff, Wales

Serra, A.
Istituto di Genetica Umana, Facultà die Medicina "A. Gemelli", UCSC, Rome, Italy

Srám R. J.
Psychiatric Research Institute, 181 03 Prague 8, Czechoslovakia

Verma, I. C.
Genetics Unit, Department of Pediatrics, All India Institute of Medical Sciences, New Delhi 110029, USA

Weppner, G.
St. Cadoc's Hospital, Caerleon, Gwent, Wales

Wertz, D. C.
Boston University, School of Public Health, 80 East Concord Street, Boston, MA 02118, USA

Ethical Issues in Molecular Genetics

K. Berg

Introduction

Ethical issues in molecular genetics as applied to man result from the potentials to interfere with or diagnose diseases, to presymptomatically detect genes for monogenic disorders, or to uncover (early in life) genetic predisposition to common disorders. The very fact that work is being conducted at the level of the gene has created not only hopes and excitement, but also a great deal of anxiety and overreaction. Some of the scenarios that create anxiety are highly fictional, and scientists have a formidable task in educating the public. There is a need to separate fact from fiction in the public debate and to aim at truly balanced, ethical decisions concerning present facts and reasonably predictable future developments. Although the extreme view that all uses of DNA technology should be forbidden may be heard, most people agree that the new gene technology offers possibilities of great importance. However, a problem in many people's minds is to find methods to safeguard against misuse of the new possibilities while securing favorable uses. To achieve a balanced development, there is an urgent need to develop well-functioning and realistic regulatory systems. In the following chapter, some of the areas of molecular biology that may create problems will be discussed. Also, regulatory systems and the present basis for ethical decision-taking will be commented on. This discussion will be limited to areas of relevance to man.

Use of Gene Technology to Produce Pharmaceuticals

The use of recombinant DNA technology to produce pharmaceuticals in microorganisms is considered by most people to be important progress that is advantageous to many patients. This application of gene technology by itself makes the position that there should be a general ban on recombinant DNA work ethically untenable. The new DNA technology has the potential for producing, in bacteria, components that would otherwise be prepared from pools of human (or animal) blood (or tissue). Even with excellent screening programs, there is no absolute safeguard against occasional blood units from infected people being included in a plasma pool that will be used to produce pharmaceuticals. The risk of causing the disease AIDS by employing pharma-

Srám, Bulyzhenkov, Prilipko, Christen (Eds.)
Ethical Issues of Molecular Genetics in Psychiatry

ceuticals prepared from human blood is ethically totally unacceptable when there is the alternative of safely producing the pharmaceutical in question by recombinant DNA technology.

Gene Therapy on Human Fertilized Eggs or Embryos

Foreign DNA which is introduced into a fertilized human egg to correct a genetic defect has the potential of being transmitted to future generations. There is at present no way of securing that a DNA construct introduced in this way will not activate oncogenes or do other harm as it occurs in a virtually unlimited number of combinations with other genes, as it passes through meioses, through generation after generation. There is no ethical justification for taking even the smallest risk on behalf of future generations. It is probably a valid ethical principle that future generations are entitled to receive DNA that has not been manipulated with.

Whenever the genetic basis for a disease is known in such great detail that gene therapy on fertilized eggs is feasible, it will also be possible to conduct prenatal diagnosis and prevent the birth of affected children by selective abortion. This procedure may be conducted as early as the first trimester, thus avoiding late abortion of affected fetuses. Prenatal diagnosis has consequences only for the fetus and the couple in question.

For the above reasons, gene therapy on fertilized eggs is not ethically acceptable at this time.

Pre-implantation Genetic Analysis of Fertilized Human Eggs

The situation concerning pre-implantation testing of fertilized human eggs is more complicated. In the mouse, one cell out of eight can be removed for analysis without apparent untoward consequences. The PCR technique makes it possible to conduct reliable analyses on very small samples of DNA. From animal experiments it appears unlikely that removing one or two cells from a fertilized human egg for analysis will have a damaging effect on the human being which could develop from that egg. Nevertheless, it is difficult to be certain in this matter. Amniocentesis for prenatal diagnosis of genetic defects became widely accepted only after it had been shown in extensive series that the procedure did not damage the child to be born. Even the smallest risk of causing harm to the human being that may develop from a fertilized egg would make pre-implantation genetic analysis very difficult to accept, from an ethical point of view. If it is possible to perform the diagnostic work on one or two cells from a fertilized egg, it will obviously be possible to conduct regular prenatal diagnosis where the risk to the fetus is known to be very small.

If pre-implantation genetic diagnostic work were to be proven harmless to the child to be born, it could be argued that the procedure is ethically preferable to traditional prenatal diagnosis. The reason for this is that abortions would be avoided, since the fertilized egg would presumably be introduced into the uterus only if the examination had shown that it did not have the genetic defect in question.

The procedure could be quite expensive (at the present time it would probably be based on *in vitro* fertilization) and this could by itself cause ethical problems. However, if the procedure were to be applied only in the highest risk categories, the number of examinations would be relatively low and the costs perhaps acceptable. In the final analysis, the decision to make use of pre-implantation diagnosis will probably depend on how much emphasis there is on avoiding abortions.

Gene Therapy on Somatic Cells

Although widely discussed for several years, gene therapy on somatic cells has yet to start. There have been considerable technical problems and there are very strict demands for extensive animal studies prior to experiments in man. Because of the risk of the introduced foreign DNA causing damage, there is broad agreement that only very serious diseases where no effective treatment is available are candidates for somatic cell gene therapy. The immunodeficiencies have been the focus of the debate, since the introduction of a normal gene into bone marrow stem-line cells could provide a "genetic cure" for severe immunodeficiency. However, other approaches may soon be developed.

Gene therapy on somatic cells is in principle, not very different from other new approaches to treatment of disease. There is a long tradition in medicine for introducing new therapies if there are realistic hopes for their success, and experience has shown that many categories of therapy carry some degree of risk to the patient. Clearly, even a therapeutic procedure with some risk involved could be acceptable if the alternative is early death or incapacity.

Since genetic manipulation of somatic cells only has consequences for the patient under treatment, the ethical problems are small in these cases, compared to those connected with gene therapy on fertilized human eggs. For catastrophic illnesses, gene therapy on somatic cells can hardly be rejected on ethical grounds if there is realistic hope that the therapy may help.

Transplantation of Fetal Brain Tissue

In the area of neurological disorders such as Parkinson's disease and Alzheimer's disease there seems to be a possibility that implantation into the patient's brain of fetal brain tissue can improve the clinical condition. If this procedure were proven to be helpful, there would be problems because of objections in some quarters to the use of material from aborted fetuses. It is important to scrutinize the foundations for such objections, and it is reasonable to require that those who argue for restrictions in this area come forward and formulate the principles on which their objections are based. Also in many other areas is it necessary that overriding principles are clearly formulated. This formulation would make ethical problems much more accessible to the public and scientists alike, and relevant value judgments would be easier to make.

It is hard to understand why it would be unethical to use material from an aborted fetus that can never become a human being if such use could prolong life or significantly improve the quality of life for people now alive. In my opinion, there is a great deal of unclear ethical thinking in questions relating to aborted fetuses. For example, where research on aborted fetuses is permitted it may be required that such research only take place if the woman has given her consent. Whereas this stance may be understandable in the case of spontaneous abortions, it is hard to see how it could be ethically justified in cases where the woman chooses abortion on non-medical grounds. In my opinion, rejection of one's own fetus is not an ethical basis for a right to determine the fate of the fetus after its death. If the basis for restistance to the use of material from aborted fetuses for research or therapeutical purposes is based on thinking that assigns full human dignity and rights to the fetus, then the woman would hardly have a right to decide over the fetus. If abortion is seriously considered as equivalent to the killing of a born human being, it would be beyond rational thinking to assign to the "killer" a right to decide over the "victim." One cannot have it both ways.

There is an urgent need for careful analyses in the areas relating to the use of fetal material for research or therapeutical purposes so that rational decisions can be made based on prevailing ethics in society.

Diagnostic Use of DNA Technology

In the near future, the main use of DNA technology in man will probably be in the diagnostic and prognostic area as well as in the area of predictive genetic testing.

Diagnosis of Genetic Diseases in Fetuses and Children

DNA technology is already widely used in the diagnostic work-up of patients. It has greatly improved diagnostic precision. DNA technology is being used to an increasing degree in prenatal diagnosis, and it can safely be predicted that it will continue to make a steadily increasing number of disorders detectable by prenatal diagnosis. Use of DNA technology in prenatal diagnosis does not seem to cause ethical problems, other than those already encountered and widely discussed in connection with prenatal diagnosis.

Presymptomatic Diagnosis of Mendelian Disorders

Several Mendelian disorders become clinically manifest only in middle age. One example of an extremely serious disorder of this kind is Huntington's chorea, which usually has its debut between ages 35 and 55. There is no efficient therapy for this catastrophic illness that causes a wide spectrum of neurological and psychiatric signs. Young people in families with Huntington's chorea generally do not know if they have the gene for the disease at the time when family planning would normally take place. Working closely with a nation-wide society for families with Huntington's chorea, I have repeatedly been told about the need that young people in such families feel for presymptomatic genetic diagnosis. The most frequently given reason for the need is that family members do not want to pass the disease on to their offspring. Many would want to abstain from having children if they knew that they carried the gene or, alternatively, to have prenatal diagnosis in order not to pass on the disease to offspring. Understandably, many at-risk individuals hope that the test will show that they do not carry the gene.

There is no DNA test as yet for the Huntington's chorea gene itself, but several DNA markers close to the gene can be analyzed. This means that, in many families, the segregation of the gene for the disease can be analyzed by following linked markers. It is, however, a significant problem that the phenomenon of recombination will, in a certain fraction of cases, separate the Huntington's chorea gene and the genetic marker in question. In practical terms this means that an *a priori* risk of, for example, 50 % (the most frequent situation when DNA examination is requested) will not be reduced to zero or rise to 100 % when the test has been performed, but, dependent on the information available in the family, fall to from 2–3 % to 8–9 % or rise to from 91–92 % to 97–98 %. Genetic counselling in families with such a serious disorder is difficult in the first place, and it is particularly difficult for the family members to comprehend the significance of an examination that does not give a "yes or no" answer but that changes probabilities. It may be an ethical problem that matters are not always fully understood, yet fully informed consent is assumed.

One significant problem with DNA diagnosis of Huntington's chorea is the danger of suicidal reactions to the finding that a person who originally had a 50 % risk has a much higher risk for carrying the gene for the disease. In our own program we are working closely with psychiatrists who are experts on handling catastrophic situations, but we are still worried that we may see tragic results of learning about the test results. We have started our testing on the basis of strong requests from the Huntington's Chorea Society, and we have discussed every problem we can think of in this connection with the board and members of the society. We have found that the autonomy of the people forcefully requesting DNA testing for genetic markers closely linked to the Huntington's chorea gene should be respected and should override other considerations.

Several sets of ethical guidelines, including one in my country, have been developed for DNA testing for Huntington's chorea. Emphasis is on the autonomy of the risk person, informed consent, free choices, total protection of test results, thorough genetic counselling, freedom to withdraw from the test program at any time, availability of psychologic/psychiatric services, and follow-up procedures. Most guidelines have totally excluded the testing of children and young persons who have not yet reached majority. Whereas the good intentions of such guidelines are above question, it could be argued that they do not sufficiently protect the rights of children and young persons. Clearly, a young individual who is still half a year or one year too young according to laws concerning majority could be as mature as many persons who have just reached majority. One has to keep in mind the very special situation that these young persons are in. They have often wondered about their own risk for many years before reaching majority, and it could be argued that their need for information is as legitimate as that of many other people. Possibly, some of the reasons for the reluctance to test young people who are close to majority reside in considerations concerning the legal aspects of informed consent. Worries concerning potential legal repercussions in certain countries may not be useful considerations in developing ethical rules for testing with respect to Huntington's chorea. At the very least, an effort should be made to balance the rights of children and young people against the well-intentioned desire to protect them from unpleasant information.

It is possible to make an "exclusion test" with respect to the Huntington's chorea gene. In such cases the "diagnosis" is dependent on the presence or absence, in the fetus, of one or more DNA markers close to the Huntington's chorea gene in the affected grandparent of the fetus. The rationale behind the test would be to interrupt the pregnancy whenever a fetus was shown to have a genetic marker inherited from its affected grandparent, which is known to be closely linked to the Huntington's chorea locus, a finding which would confer upon the fetus a 50 % disease risk. If in such cases the couple were to decide to continue the pregnancy (after examination had uncovered that the fetus would have a 50 % risk of developing Huntington's chorea), a particularly difficult situation could develop. If the father or mother of the fetus had a 50 % risk of developing Huntington's chorea, and the same risk was present in the child, signs of this serious disorder in the fahter or mother at risk would confirm that the developing child would contract Huntington's chorea. This development could take place years after the child had been born, but still many years before a legally valid statement of informed consent could be obtained from the child. One response to this potential problem would be to refuse to conduct prenatal "exclusion testing" with respect to Huntington's chorea. On the other hand, such an attitude would be unfair to the majority of the people requesting a prenatal "exclusion test" who would most likely want to interrupt the pregnancy if a 50 % risk were to be identified in the fetus, and who would therefore not create a situation where it would become known that a living child has the gene for Huntington's chorea. The needs of this majority of people have to override other considerations, but the possibilities outlined above underscore the importance of in-depth genetic counselling prior to DNA testing with respect to Huntington's chorea.

Predictive Genetic Testing for Disorders of Middle Age

For several common disorders, including atherosclerosis, allergies and psychoses, there is little doubt that genetic factors are of importance. This is particularly well studied in coronary heart disease.

The dominant approach to identifying the genes involved in the etiology of common disorders is the so-called candidate gene approach. For example, with respect to coronary heart disease and atherosclerosis, there are several categories of candidate genes, such as any gene whose protein product is:

1. involved in lipoprotein structure, lipoprotein metabolism or lipid metabolism
2. involved in thrombogenesis, thrombolysis or fibrinolysis
3. involved in regulation of blood flow in coronary arteries
4. involved in regulation of blood pressure
5. involved in reverse cholesterol transport
6. present in atherosclerotic lesions
7. involved in the regulation of growth of atherosclerotic lesions

The genes involved in the control of Lp(a) lipoprotein are particularly important in determining an individual's susceptibility or resistance to coronary heart disease. Genes belonging the genetic polymorphisms in apolipoprotein E, as well as DNA variants at the apolipoprotein B locus, influence the population variation in the risk factor cholesterol (for review, see Berg 1990a,b). There is realistic hope that examination of a limited number of genes will in the future help to identify the majority of people with increased genetic susceptibility to coronary heart disease.

Identification of important susceptibility genes will make it possible to detect genetic predisposition to coronary heart disease at a young age. This will give persons at risk the option of embarking upon very active preventative actions from early in life. The people who learn that they have a specific risk of contracting coronary heart desease will probably be much more strongly motivated toward a healthy lifestyle than the population at large, and the combination of strong motivation and an early start of preventative efforts holds considerable promise for improved disease prevention.

Although there is little doubt that genes are important in the etiology of psychiatric illnesses and in determining behavior (Cloninger 1990; Guze 1990; Kennedy et al. 1990; Kendler 1990; Plomin 1990; Mohr 1990), the number of candidate genes appears to be so high that it may be a long time before a reasonable number of them are identified. The identification of single genes contributing to the risk for psychiatric disorders could greatly improve the understanding of such conditions. For the time being there are no preventative procedures available to at-risk persons, as there are with respect to coronary heart disease. This has consequences regarding whether or not screening with respect to disease susceptibility would be desirable (see below).

Although screening with respect to genes conferring risk for coronary heart disease or other frequent disorders could be conducted to prevent disease, results of predictive genetic testing could be used to an individual's disadvantage. Clearly, the chance of promotion to a highly responsible positions could be reduced if it were known that a candidate had an increased risk of suddenly dying or becoming incapacitated. It could be even worse if it were known (from psychological/psychiatric predictive testing) that a person might not handle crises in an optimal manner. Also, insurance companies could be tempted to use predictive genetic testing as part of risk and premium estimation, at least for very high insurance policies.

An Area in Need of Legislation

Surprisingly enough, relatively little attention has been paid to the serious problems of data protection in connection with predictive genetic testing and screening. Existing traditions for handling data about previous or present diseases are, in my opinion, not adequate for the handling of results of predictive genetic testing.

It is my conviction that, whenever genetic observations are made in persons (or fetuses) who are healthy at the time of examination, the resulting information should *not* be handled in the same way as clinical information on overt disease. Information on manifest diseases is traditionally requested by and given to life insurance companies, employers, pension funds, educational institutions, the military and other consumers of health data.

Information concerning a healthy person's genes should, in my opinion, be considered as the property of the person. The law should explicitly forbid employers, life insurance companies and other consumers of health information to even ask if testing for single genes causing disease or predictive genetic testing for disorders of middle age has been performed. This appears to be the only way to secure safe, beneficial use of predictive genetic testing and screening to combat disease. It would be highly desirable if society gave priority to securing safe handling of such data. This being achieved, most of the problems in connection with genetic predictive testing and screening would be easier to solve.

Genetic Screening

Screening is the examination of a relatively large section of a population or subpopulation in the absence of a suspicion that any specific person or fetus may have a disorder, disease risk, or chance of transmitting genetic risk to offspring. This is distinct from "diagnosis", which refers to an examination conducted to verify or reject a specific suspicion of disease in a given person or fetus. Screening of various kinds has taken place in many countries for decades, and the purposes have been to prevent disease or to secure early treatment in an individual, to protect the society against the spread of a disease, or to arrive at new knowledge.

The progress in molecular genetics has greatly increased the potential for screening with respect to genetic diseases or predisposition to disease. This development has caused considerable anxiety, and some national and international organizations, including the Council of Europe through its committee CAHBI, are developing guidelines for genetic screening. Because of the potential that well-planned screening studies have for increasing knowledge and for improving disease prevention, it is important that public anxiety does not lead to irrational decisions concerning screening. I think that the good offices of WHO have a specific responsibility to make certain that the health aspects are in the foreground in these discussions, and to help to separate fact and fiction.

In my opinion, screening for genetic disease, heterozygous carrier state or genetic predisposition to common diseases is ethically acceptable if:

1. the screening leads to important new understanding of the disease
2. the test results can be used to promote health
3. the test results can be helpful for reproductive choices or other planning for people at high risk of contracting or transmitting a serious genetic disorder
4. screening is voluntary
5. test results are so strictly protected that they cannot be used to the disadvantage of the tested individual or the person's family
6. adequate information about the test is given prior to testing
7. comprehensive genetic counselling is offered whenever an adverse test result is encountered

It would be useful if WHO would lend its prestige and knowledge to decide to what extent guidelines and rules for screening are needed, and would develop such rules for predictive genetic testing that it might find useful.

It would also be very useful if WHO would conduct discussions, at certain intervals, concerning diseases that are candidates for screening at any given time. Newborn screening or screening of pregnant women are realistic possibilities with respect to cystic fibrosis and Duchenne muscular dystrophy, and the results of such tests could be of great importance for future reproductive choices and could give gene carriers the option of having prenatal diagnosis conducted in later pregnancies. Testing for susceptibility to juvenile diabetes may become practically important because of the potential to interfere with the autoimmune process when there is still time.

Screening for diseases of adult life that can be prevented, such as coronary heart disease, should be encouraged, and a family-oriented preventive medicine should be developed to help individuals and families cope with disease risks, whenever detected.

Do We Have the Instruments to Handle Ethical Problems in Molecular Biology with Respect to Man?

A frequent response of several societies to public worry it to create lay committees that are meant to protect society's interest. In the area of biomedicine, there is today a rich tapestry of ethics committees, and the power of such committees is often very real, even though in many instances they formally, serve only in an advisory capacity. They have become a major factor in everyday life in biomedicine and one's success or failure with ethics committees by decide the fate of a research project, influence research careers, affect recruitment to biomedicine and, in the long run, decide whether or not a given area of research is destined to survive.

In some countries, ethics review committees that deal with biomedical problems may have very few scientists or medical doctors among their members and may consist mainly of lawyers, philosophers and theologians in addition to lay people. The non-expert members of the committees may be recruited among media personalities, frequently among those who have publicly expressed concern about activities in biomedicine.

Attitudes encountered in some ethics review committees, even some that have only an advisory capacity, are such that the committee's ego appears to be that of a court. Decisions handed down may be phrased in "court language" or expressed as orders rather than advice. The individual scientist has very little chance to get around a decision or order by the committee, and the consequences of such decisions make it reasonable to consider some ethics review committees as courts.

People who have been trained by society for a great number of years to qualify as high level scientists have a right to conduct the work they have been trained for. A denial of this right is indeed a serious interference with the individual's freedom. Such interference with the right to do one's work can easily stand a comparison to consequences of decisions made by courts.

It is not necessarily bad to have ethics review committees with powers resembling those of courts. However, for such a system to be acceptable it must fulfill several criteria for good court practices in democratic societies. Thus, in a given country all courts must follow the same law (and all ethics committees should follow the same set of principles and rules), decisions should be appealable, professional judges should be experts whose decisions are widely respected (the equivalent in ethics review committees would be a future breed of highly competent experts in bioethics) and, if lay jurors are to be employed, they should be truly representative of the population, not chosen because they have expressed skepticism towards "the accused" (or towards a biomedical community which the person under examination represents).

There is considerable uncertainty with respect to the basis for decisions made by ethics review committees. Some guidance has been sought in existing sets of rules for biomedical research, and the declaration of Helsinki is often referred to. This document's emphasis on "therapeutically oriented" research may not be totally adequate as a guideline for work in molecular biology with respect to man. Although it is not stated in the declaration of Helsinki, it is sometimes implied that this document requires research on human beings to be of direct advantage to those studied. This is an extremely short-sighted attitude. There is a need to introduce in the thinking around research ethics and informed consent an element of group solidarity and duty. The one-sided emphasis on the individual's *rights* needs to be supplemented with statements concerning human *duties* and obligations. One of those duties should be to participate in research if it can be to the advantage of one's own group, even though the advantage to the individual under study may be negligible or nonexistent.

In some quarters there appears to be some confidence that "bioethics" will provide a solid basis for the work of ethics review committees. The belief in this concept is so strong that some initiatives have already been taken to start work on a European convention on bioethics. I believe that a possible future convention should have a broader perspective, and I see important roles for WHO in this area.

In the concluding session of the First European Congress of Bioethics organized by the Council of Europe in Strasbourg, December 5–7, 1989, three different definitions of "bioethics" were used by three different key speakers. The fact that the best brains in "bioethics" have not yet arrived at a definition of the field makes it difficult to accept that it can at the present time provide useful guidelines for regulation of activities in molecular genetics with respect to man.

Concluding Remarks

An operational definition of "bioethics" as relating to man should include that it is a system of consistent thinking aimed at resolving conflicts between, on the one hand, (usually legitimate) needs with respect to research, patient treatment or disease prevention and, on the other hand, the need or desire to protect people from harm or significant discomfort. Guiding principles in a new "bioethics" could be that research activities are accepted as important if they can lead to beneficial consequences and that people who have been trained by society to conduct scientific research have a right to do so unless they could cause harm or significant discomfort to patients or other persons. WHO should take the lead in developing the field of bioethics.

References

Berg K (1990a) Level genes and variability genes in the etiology of hyperlipidemia and atherosclerosis. In: Berg K, Retterstøl N, Refsum S (eds) From phenotype to gene in common disorders. Munksgaard, Copenhagen, pp 77–91

Berg K (1990b) Genetics of atherogenic lp(a) lipoprotein. In: Berg K, Retterstøl N, Refsum S (eds) From phenotype to gene in common disorders. Munksgaard, Copenhagen, pp 138–162

Cloninger CR (1990) Gene-environment interaction in the development of personality and its disorders. In: Berg K, Retterstøl N, Refsum S (eds) From phenotype to gene in common disorders. Munksgaard, Copenhagen (in press)

Guze SB (1990) Genetic aspects of alcoholsim. An update. In: Berg K, Retterstøl N, Refsum S (eds) From phenotype to gene in common disorders. Munksgaard, Copenhagen (in press)

Kendler KS (1990) Schizophrenia as a complex phenotype. In: Berg K, Retterstøl N, Refsum S (eds) From phenotype to gene in common disorders. Munksgaard, Copenhagen (in press)

Kennedy JL, Giuffra LA, Cavalli-Sforza LL, Pakstis AJ, Kidd JR, Sjøgren B, Wettersberg L, Kidd KK (1990) Searching for genes predisposing to neuropsyciatric disorders. In: Berg K, Retterstøl N. Refsum S (eds) From phenotype to gene in common disorders. Munksgaard, Copenhagen (in press)

Mohr J (1990) Genetic aspects of Alzheimer's disease. In: Berg K, Retterstøl N, Refsum S (eds) From phenotype to gene in common disorders. Munksgaard, Copenhagen (in press)

Plomin R (1990) The role of inheritance in behavior. Science 248: 183–188

A Single Locus for Psychosis and Intelligence in the Exchange Region of the Sex Chromosomes?

T. J. Crow

Since Kraepelin distinguished the typical features of manic-depressive illness from those of schizophrenia it has been widely assumed that these are two separate diseases. Prevailing opinion is that there will be more than one genetic locus for affective illness (at least one being autosomal) and one or more (autosomal) loci for schizophrenia. Favoured candidate genes are those coding for neurotransmitter receptors and enzymes.

The contrary view, to be presented here, is that there is but one locus and that multiple alleles account for the variety of affective and schizophrenic forms of psychotic illness. The candidate gene is the "right shift factor" or cerebral dominance gene; the hypothesis is that this locus is not autosomal but within the exchange (or pseudoautosomal) region of the sex chromosomes.

The Continuum Concept

Emil Kraepelin's key contribution was to establish that there is a relationship between the form of a psychotic illness and its outcome. Illnesses in which affective change is a major or prominent component on the whole have a better outcome than those in which affective features are less prominent and in which the psychotic phenomena cannot be understood as secondary to affective change. The success of the Kraepelinian system, even its survival, owes everything to the reality of this relationship. Yet the independence of the two Kraepelinian entities has never been satisfactorily established.

Doubts about the Kraepelinian binary concept were first raised by Kendell and Gourlay's (1970) study of the patient population in the US-UK diagnostic investigation.

They devised a discriminant function (Fig. 1) that achieved maximum separation of the populations of patients with a project diagnosis of affective disorder from those with a project diagnosis of schizophrenia: items favouring affective illness (early morning waking, delusions of guilt and so forth) counting to the right, and items favouring a diagnosis of schizophrenia (Schneiderian first rank features, affective flattening) counting to the left. The prediction of the two entities principle is that there will be a bimodal distribution, with the major entities at either end and relatively few cases in the middle. The finding is that there is a peak at the mid-point, i.e., that schizo-affective cases are common

Srám, Bulyzhenkov, Prilipko, Christen (Eds.)
Ethical Issues of Molecular Genetics in Psychiatry

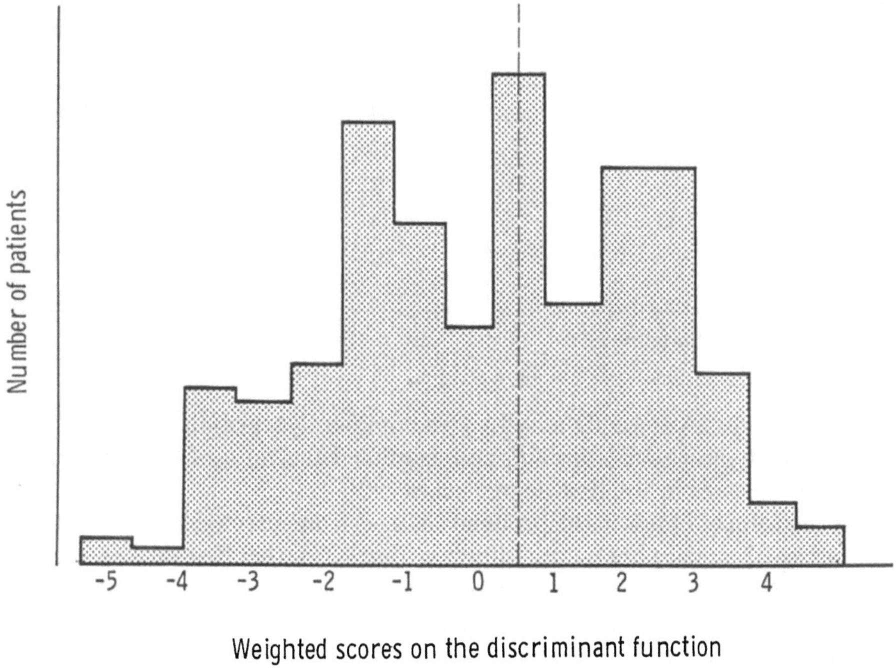

Fig. 1. The distribution of weighted scores of 146 patients with a diagnosis of schizophrenia and 146 patients with a diagnosis of affective disorder from the US-UK study on the discriminent function derived by Kendell and Gourlay (1970) to maximise separation of the conditions

rather than relatively rare, as is required by the binary concept. Kendell and Gourlay (1970) repeated the investigation on a second sample, on this occasion obtaining a unimodal rather than a trimodal distribution. The investigation thus posed a serious challenge to the Kraepelinian viewpoint that has never been answered.

Kraepelin (1920) himself was aware of the problem. In one of his last contributions he wrote:

> "Perhaps it is also possible to tackle the difficulties which prevent us from distinguishing reliably between manic-depressive insanity and dementia praecox. No experienced psychiatrist will deny that there is an alarmingly large number of cases in which it seem impossible, in spite of the most careful observation, to make a firm diagnosis ..... Nevertheless it is becoming increasingly clear that we cannot distinguish satisfactorily between these two illnesses and this brings home the suspicion that our formulation of the problem may be incorrect."

Recent genetic evidence supports a continuum rather than two entities. For example Angst and Scharfetter (1990; see Angst et al. 1983 for an earlier report on these studies) studied over 250 probands with diagnoses covering the continuum of psychotic illness as it extends from unipolar depression (UP) through bipolar affective (BP) disorder, schizo-affective illness of predominant-

Table 1. Morbid Risk in First-Degree Relatives

	UP	BP	ASA	SSA	S
Probands	58	31	34	35	105
Morbid risk: Schizophrenia (S)	10	4	19	18	31
Affective psychoses (A)	24	6	14	4	5
S : A ratio (age-corrected)	0.30	0.47	0.92	2.99	5.05

From Angst and Scharfetter 1990

ly affective type (ASA) and then (S) of predominantly schizophrenic type (SSA), through to typical schizophrenia, and examined the ratio of schizophrenic to affective illnesses in the first-degree relatives.

As the form of illness in the proband changes from affective to schizophrenic, the ratio of schizophrenia to affective illness in the first-degree relatives increases. There is no discontinuity such as would allow one to assert that there are two quite separate genetic components. The conclusions of Angst and his colleagues from their recent studies (Stassen et al. 1988) are that:

> "Typical syndrome patterns appeared in both populations (i.e., both probands and relatives) ..... Nevertheless our analyses revealed no clear breeding true of either affective disorder or schizophrenia".

Other studies are in agreement. Thus Ødegaard's (1972) investigation yielded conclusions which are almost identical to those of the Zurich group. Of particular interest are the successive contributions of E.S. Gershon and colleagues at NIMH. Gershon et al. (1982) concluded that schizoaffective psychoses were the furthest extremity of the continuum of affective disorders, and then when they started studying probands with schizophrenia (Gershon et al. 1988), these workers found that schizo-affective illness was also genetically related to schizophrenia. The conclusion is inescapable that there is a continuum rather than two separate entities. Kraepelin was right to insist that there are proto-typical affective and schizophrenic illnesses and that form of illness predicts outcome, but wrong to assert that there are two aetiologically distinct diseases. The recent studies of Maier and Hallmeyer (Maier et al. 1989) in Mainz reinforce the continuum concept.

A problem with many studies which apparently have established the genetic independence of the two proto-typical psychoses is that they have used samples from which the intermediate and atypical psychoses are eliminated. A notable illustration of the bias introduced by sampling, as probands only, those patients who suffer from classical affective and schizophrenic psychoses is provided by Kendler et al. (1986). Whereas previously Kendler et al. (1986) had applied DSM III criteria for affective disorder and schizophrenia to the Iowa 500 series of patients and controls and their relatives, and had concluded that the Kraepelinian separation was generally supported by the findings in the relatives, in this study they examined a series of probands with a chart diagnosis of schizophrenia who did not meet the Feighner criteria. These individuals were allotted to the DSM III diagnostic categories of schizophreniform disorder, schizo-affective disorder and psychotic affective illness. Kendler et al.'s (1986) conclusions deserve full quotation:

> "The pattern of psychopathology in relatives of schizophreniform probands closely resembled that found previously in relatives of schizophrenic probands. Relatives of schizo-affective probands had an excess risk for schizophrenia, other psychoses, and bipolar illness. The pattern of illness found in relatives of probands meeting Research Diagnostic Criteria for mainly schizophrenic schizoaffective disorder appeared indistinguishable from that of relatives of schizophrenic probands. Relatives of probands with psychotic affective disorder had an excess risk for schizophrenia and for unipolar and bipolar affective disorder."

It is instructive to examine these conclusions in an attempt to determine where the dividing line between schizophrenia and affective disorder might be drawn. First since the spectrum of symptoms in the schizophreniform disorders is the same as in schizophrenia, while their duration is less, the similarity in genetic relationships is to be expected. Secondly the finding that the relatives of patients with schizo-affective psychoses have an excess risk for both schizophrenia and affective illness is problematic for the Kraepelinian viewpoint, unless these psychoses can be subdivided into schizophrenic and affective subtypes. But this is just what the studies of Angst et al. (1983) have failed to establish; no such distinction can be justified. Lastly the findings in the relatives of patients with psychotic affective disorder in this series suggest a greater overlap between the proto-typical psychoses than was apparent in the earlier DSM III based studies on the Iowa series.

Overall the findings give no comfort to the Kraepelinian concept. The impact on Kendler's own thinking is interesting to follow. Previously, in collaboration with Hays (Kendler and Hays 1983), he had studied patients with schizophrenia defined by DSM III criteria and had identified a group of 28 with a first-degree relative with affective disorder. These were compared with a group of 98 with no history of affective disorder in a first-degree relative. Patients with an affective family history were more likely to have suffered from affective symptoms on follow-up. The conclusion was that "even when DSM III criteria are met hesitation is indicated in diagnosing schizophrenia in patients with a first degree relative with bipolar illness."

In the later paper, however, Kendler et al. (1986) conclude that: "The familial link between schizophrenia and at least a sub-group of schizo-affective disorders argues that the presence of prominent affective features, both depressive and manic, during the course of a chronic 'schizophrenic-like' illness need not indicate a familial predisposition to affective illness".

One must ask the defenders of the Kraepelinian concept the question, where is the line to be drawn? Is it to be drawn so as to define schizophrenia restrictively as it seems that Kendler and Hays in 1983 would have recommended, or is it to be drawn to include a substantial fraction of schizo-affective disorders, as Kendler et al. in 1986 suggest? In this case we are faced with the problem of distinguishing the "affective" types of schizo-affective disorder from the "schizophrenic" types. But this is just what the investigations of Angst et al. (1983; and Angst and Scharfetter 1990) and Gershon et al. (1982, 1988) suggest cannot be achieved. The very existence of these illnesses is a challenge to the two entities principle, as Kraepelin himself appreciated. Their genetic relationships, as demonstrated by the recent studies, deal a death-blow to the Kraepelinian system.

Thus it is difficult to identify a coherent defence in the literature of the classical Kraepelinian position. It must be assumed that that position has given way to some form of the continuum hypothesis: either that there are a large number of different genetic entities, each with its own locus, that show substantial overlap in their phenotypic manifestations (Kendell 1988), or that there are a number of different alleles at a single genetic locus (Crow 1986a, 1987). The first view has difficulty in explaining the extent and source of variation within families. The second faces the problem of the nature of the variation at the hypothetical locus. What sort of genes are these that show an apparently continuous variation? What sort of change in the gene could take place between different members of the same family?

As I previously suggested (Crow 1986a):

> ".... the genes predisposing to psychosis have a degree of stability that ensures that the form of psychosis tends to remain the same within families, but there is also the possibility of change, implying that the genetic mechanisms themselves are variable".

The genes are changing, but in small steps and in a coherent way. The only plausible model is that we have a number, perhaps a large number of alleles at a single genetic locus.

Implications for a Locus

Where is the gene? In October 1987 the *Lancet* discussed the alternative to Kraepelin's binary viewpoint – the continuum concept of psychosis – in a sympathetic but penetrating way (Leader 1987), emphasising the undoubted corollary, that there should be a single genetic locus. Yet quite apart from the various claims to have established linkage for one or more of the psychoses, we have reasonable evidence that the affective disorders at least are sometimes sex-linked (Winokur et al. 1969) and sometimes not (Rice et al. 1987). Does this not indicate that heterogeneity is already established?

My response (Crow 1987) was that there remained one way in which homogeneity could yet be maintained. This was that the gene for psychosis should be located within the pseudoautosomal (or exchange) region of the sex chromosomes. The pseudoautosomal region is that small region (described in more detail below) of the short arms within which there is exchange of genetic material between X and Y chromosomes in male meiosis. Such a locus allows us to see how there might be a form of sex linkage in a condition which generally follows an autosomal mode of inheritance. To maintain this view it is necessary to regard existing claims for autosomal linkage – for example, to chromosome 11 (Egeland et al. 1987) – as unfounded, and that now appears more plausible (Kelsoe et al. 1989) than it did two years ago.

There are other reasons for taking a sex chromosomal locus for psychosis seriously. Forssman (1970) discussed the excess of cases of sex chromosome aneuploidy – particularly XXY and XXX – that have consistently been observed in populations of patients with psychosis. He wrote:

> "Myself I am convinced that study of aberrations of the sex chromosomes will prove to have a radical effect on large sections of psychiatric thinking" and "That several different types of gonosomal aberration are associated with an increased risk of functional psychosis is an observation which should make us stop and think".

We have two alternatives. Either we suppose that the psychoses associated with sex chromosome aneuploidies represent an atypical and genetically distinct form of schizophrenia, and that the rest are as has always been supposed related to a gene or genes somewhere on the autosomes, or we can conjecture (Crow 1987, 1988) that the locus for psychosis is, contrary to what has generally been supposed, on the sex chromosomes. That a locus on the sex chromosomes may be combined with an autosomal pattern of inheritance comes from our recently acquired knowledge of the pseudoautosomal region.

Figure 2 is taken from a paper by Burgoyne (1986) who introduced the term pseudoautosomal. The possibility of exchange between X and Y chromosomes was first suggested by Koller and Darlington (1934). That it actually takes place is established by molecular studies such as those of Cooke et al. (1985), Goodfellow et al. (1986), and Weissenbach et al. (1987). Within the region there is strict homology of genes on the two sex chromosomes, although outside it there is divergence, including at the testis-determining factor that establishes sex. Genes within the region can be transmitted from father to son as well as from father to daughter; hence the pattern of transmission is autosomal. Because a single obligatory cross-over occurs in male meiosis, the rate of recombination here is high relative to that elsewhere in the human genome, and there may also be a high rate of unequal recombination – a possible mechanism of generation of new mutations. The characteristic feature of pseudoautosomal transmission is same sex concordance, that is, that affected siblings will be more often than would be expected of the same sex. This arises because when the gene travels from father to child it may be carried on the X chromosome to daughters, or on the Y chromosome to sons; it could be either, but affected children are more likely to be of the same sex.

The presence of same sex concordance in psychosis was first noted by Mott (1910), and later discussed by Penrose (1942) and Rosenthal (1962), but none of them could have been aware of the pseudoautosomal explanation. We have

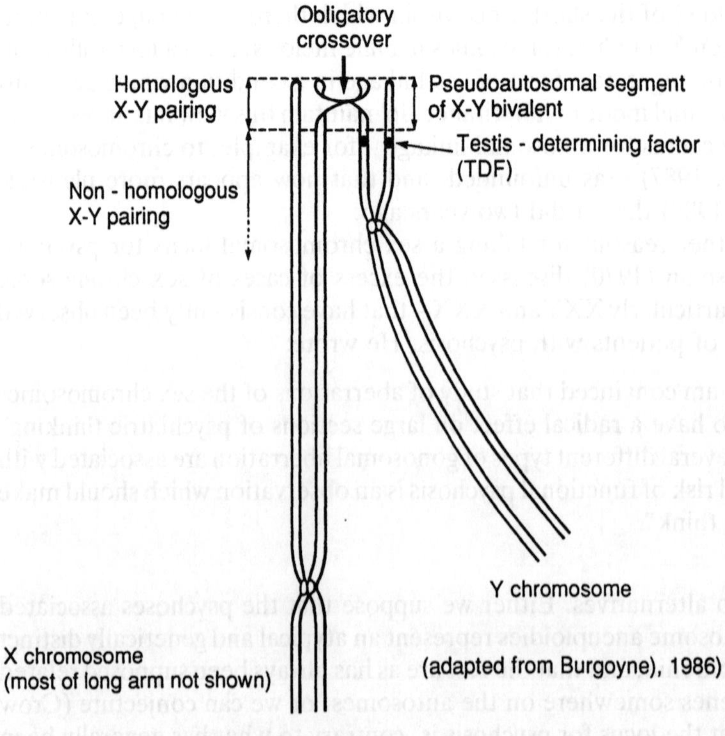

Fig. 2. Crossing over between X and Y chromosomes in male meiosis in the human pseudoautosomal region. (From Burgoyne 1986)

presented evidence (Crow et al. 1989c) that, as predicted by the pseudoautosomal hypothesis, such concordance is associated with paternal transmission. This interpretation was challenged by Curtis and Gurling (1990) on the grounds that we had failed to take proper account of the excess of males in our series. When allowance was made for the male bias, same sex concordance and an association with paternal transmission remained (Crow et al. 1990).

In addition there is molecular evidence for linkage to a pseudoautosomal probe from a study (Collinge et al. 1989) on a subset of sibling pairs. The telomeric probe DXYS14 is unlinked to sex and gives rise to a polymorphic series of bands on Southern blots that is suitable for an affected sibling pair analysis. This approach makes no assumptions about the mode of transmission, whether recessive or dominant, and does not depend upon allocating to relatives diagnoses whose status, e.g., as borderline states, may be uncertain. Sibling pairs are selected on the basis that they are both definitely affected, and one then examines whether at the given locus they share alleles from the parents at or above chance expectation. If the parents have alleles *ab* and *cd* the possibilities in the children are that they share both alleles – say *ac* – from the parents giving an IBD score of 2, that they share one allele – say *a* – giving an IBD score of 1, or that they share no alleles, with an IBD score of 0. The chance expectations of these outcomes are in the ratio of 1 to 2 to 1.

In 68 pairs of siblings, of which we would expect 17 sibling pairs to share two alleles, we find that 22 have done so, and where we would expect 17 to share no alleles, we find that only 11 have done so. This is a departure from chance expectation significant at the 2 % level, although inclusion of multiple sibships in the same sample reduces the significance to 0.036. All that we have done in selecting these siblings is to ensure that both suffer from either schizophrenia or schizoaffective disorder. Such siblings will share a piece of chromosome at the psychosis locus, i.e., they will share that part of the chromosome because they have inherited it from an affected or carrier parent. If the gene is recessive they will share the relevant part of the chromosome from both parents. However, selecting siblings for psychosis should have no influence on the distribution of alleles unless the locus in question is close to, i.e., linked to, the psychosis locus. The result, therefore, supports (although weakly) the view that there is a linkage between schizophrenia and this genetic locus at the telomere of the short arms of the X and Y chromosomes. Further support for the hypothesis comes from the observation that the shared alleles are significantly more likely to have come from the mother than the father, a finding that is predicted on the basis that recombination within the region is 10 to 20 times more frequent in male than in female meiosis.

If the continuum concept holds, same sex concordance should also be present in affective illness. There is evidence that it is (Table 2).

Table 2. Distribution by Sex in Sibling Pairs with Bipolar Affective Illness[a]

BP pairs only:	Winokur et al. (1969)	Goetzl et al. (1974)	Mendlewicz and Rainer (1974)	Gershon et al. (1982)	Totals
MM	3	0	19	7	29
MF	4	0	19	15	38
FF	5	3	21	9	38
Same Sex	67				
Opposite Sex	38		chi-square = 8.4 $p<0.01$		

[a]Adapted from Crow (1988). M, male; F, female.

As in schizophrenia, concordance by sex is present, although no information on whether it is associated with paternal transmission is available. The findings of the National Institute of Mental Health Collaborative Program on the Psychobiology of Depression with respect to familial transmission of bipolar illness (Rice et al. 1987) are also relevant. In this study of 187 families of bipolar patients, risk to relative was found to be unrelated to sex of proband or relative. However, there was heterogeneity in sex-specific sibling correlations, in that brother-sister correlations (0.36 for interview and 0.23 for history data) were significantly ($p<0.05$ and $p<0.005$, respectively) less than brother-brother (0.63 and 0.53) and sister-sister (0.53 and 0.54) correlations. Such correlations may reflect a locus in the pseudoautosomal region.

What Does the Gene Do?

That there are structural changes in the brain in schizophrenia was first suggested by pneumoencephalographic (e.g., Haug 1962) and computerised tomographic studies (Johnstone et al. 1976). There is a consensus that in those studies, which include reasonably large samples of patients with the more chronic forms of disease, there is a modest degree of enlargement of the lateral cerebral ventricles. Whether this enlargement precedes onset of illness and/or increases with disease progression is unclear.

Two post-mortem studies have clarified the nature of the changes. In the first (Brown et al. 1986), the brains of 41 patients with schizophrenia (diagnosed according to the Feighner criteria) were compared with those of 29 patients with affective disorder. The brains of patients with schizophrenia:

1. were 6 % lighter;
2. had lateral ventricles that were larger in the anterior (by 19 %) and particularly the temporal (by 97 %) horn cross section; and
3. had significantly thinner parahippocampal cortices.

The findings provided confirmation of neuroradiological studies and suggested that the structural changes might relate particularly to the temporal lobe. In the parahippocampal gyrus there was a significant diagnosis by side interaction, the difference between the groups being greater on the left side of the brain.

The second study (Crow et al. 1989a) reinforces the emphasis on laterality. Brains of patients with schizophrenia and normal controls were collected prospectively and the structure of the ventricle was assessed on X-ray images after infusion of a radio-opaque dye. In this way the extent of the ventricle could be determined in the lateral aspect. Again enlargement was greatest in the posterior and particularly the temporal horns; in addition it was highly selective to the left hemisphere (Fig. 3). The finding was true for brains of both males and females, and was unaffected by the presence of histopathological abnormalities, e.g., cerebrovascular disease or Alzheimer-type change. In brains of patients with Alzheimer-type dementia, assessed in parallel with the schizophrenic series, temporal horn enlargement was also present, but this showed no selectivity for the left side of the brain.

Some recent radiological studies also have revealed lateralised changes. Thus in a CT scan study of discordant MZ twins, Reveley et al. (1987) found reduced CT scan density (perhaps reflecting increased cerebrospinal fluid spaces) on the left but not on the right side in the affected twin. In a CT scan study of chronic patients, the width of the posterior half of the brain on the left side was reduced in patients with an early onset of illness (before 25 years for males and 28 years for females) by comparison with those of later onset of illness (Crow et al. 1989b), and in an NMR comparison of patients with affective disorder and schizophrenia a significant diagnosis by side interaction was observed for temporal lobe area, with a relative reduction on the left side in the schizophrenic patients (Johnstone et al. 1989).

Each of these findings is consistent with the concept that the changes in the brain are lateralised. The findings with respect to temporal horn area in the second post-mortem study (Crow et al. 1989a) suggest that such lateralisation is specific to schizophrenia. One possibility is that there is an arrest of development of the left temporal lobe; such a change might account for a reduction in width and area of the temporal lobe, as seen in the neuroradiological studies, and also for the increase in temporal horn area seen in the post-mortem study.

The notion that schizophrenia is a disease of cerebral asymmetry owes much in recent years to the interest generated by Flor-Henry's report that the psychoses

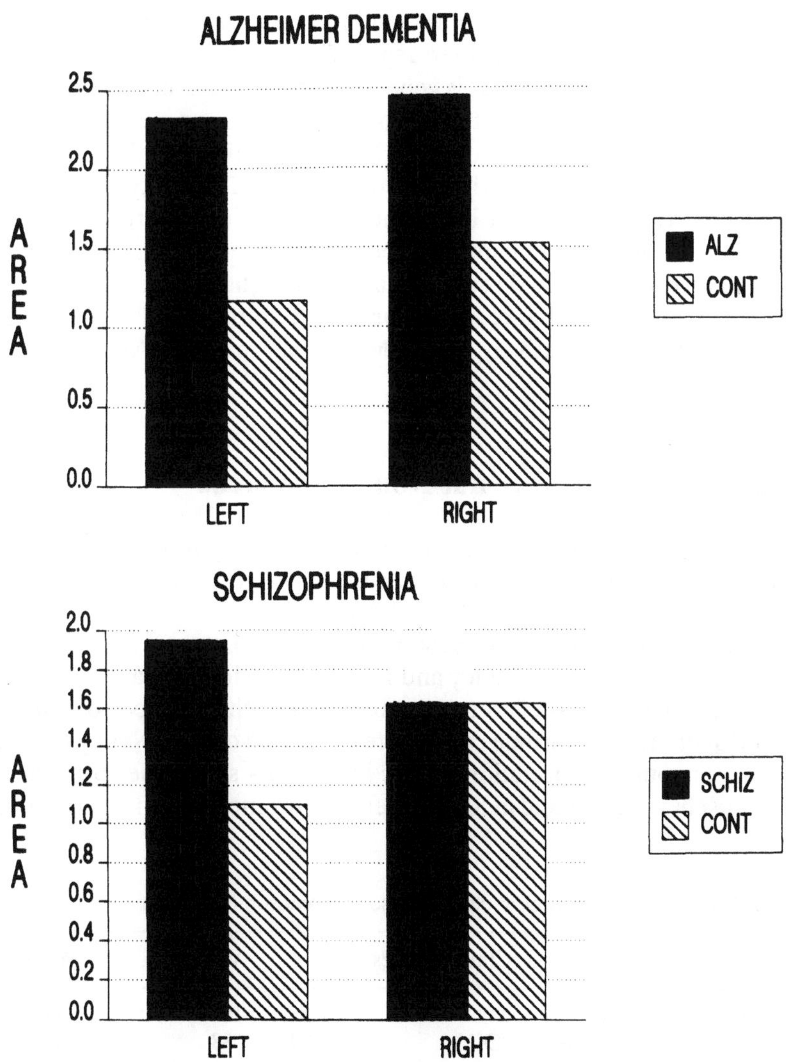

Fig. 3. Temporal horn area (assessed from the lateral aspect) in schizophrenia and Alzheimer-type dementia (black columns) by comparison with age-matched controls (cross-hatched). By ANOVA there is a significant diagnosis by side interaction ($p < 0.001$). (From Crow et al. 1989a)

associated with temporal lobe epilepsy are schizophrenia-like when the focus is on the left side of the brain. But this notion has a history which extends back into the last century. Thus in 1879 Crichton-Browne wrote that "the cortical centres which are last organised, which are the most highly evolved and voluntary, and which are supposed to be located on the left side of the brain, might suffer first in insanity". In 1881 Luys assessed the weights of the two halves of the brain and claimed that, whereas in normals the left side of the brain is generally heavier, in patients with mental illness the opposite is often the case. Southard in 1915 summarised his own studies of the neuropathology of schizophrenia as showing that "the atrophies and aplasias when focal show a tendency to occur in the left cerebral hemisphere" and added that "for this there is probably good a priori reason in the structure, late evolutionary development, and consequent relatively high lability of these regions".

How could a disease that is genetic in origin cause an asymmetric distribution of pathological change in the brain? The obvious answer is that the genes that are subject to variation in disease themselves influence the development of the asymmetries in the human brain. Geschwind and Levitsky (1968) have described asymmetries in temporal lobe that relate particularly to auditory association cortex, including the planum temporale. The genes that control these developments presumably are those which also control handedness. This genetic factor – the "cerebral dominance gene" or "right shift factor" of Annett – may be transmitted in a relatively simple Mendelian manner (Annett 1985; McManus 1985).

Location of the Cerebral Dominance Gene

A clue to the location of the cerebral dominance gene comes from the studies of the neuropsychology of Turner's and Klinefelter's syndromes that have been conducted by Netley and Rovet (1982, 1987).

Turner's syndrome cases lack an X chromosome and Klinefelter's syndrome cases have an extra X. Netley and Rovet (1982) found that on IQ tests they have reciprocal deficits – in Turner's syndrome cases there is a performance deficit and in Klinefelter's syndrome cases there is a verbal deficit (Netley and Rovet 1987). The obvious interpretation is that in Turner's syndrome cases there is a right hemisphere deficit and in Klinefelter's syndrome cases there is a left hemisphere

Table 3. Neuropsychological deficits in sex chromosome aneuploidies

	Chromosomes	n	Verbal IQ	Performance IQ
Turner's syndrome	XO	35	100	88
Klinefelter's syndrome	XXY	24	83	100

(data taken from Netley and Rovet, 1982, 1987)

[a] Note: in a collated sample of XXX girls a verbal-performance discrepancy similar to, but of lesser magnitude than, that in XXY males was seen (Netley 1986)

deficit. Thus some factor located on the X chromosome is responsible for the relative development of the two hemispheres. However, because in normal females one X chromosome is inactivated, the abnormalities in Turner's syndrome cases are attributable to those parts of the X that are not inactivated (Ferguson-Smith 1965). These include particularly the pseudoautosomal or exchange region. This suggests that the cerebral dominance gene is in the pseudoautosomal region or some other part of the X chromosome that is not inactivated (Crow 1989).

Evolutionary Origins

What is the evolutionary significance of the cerebral dominance gene?

In his book entitled Evolution of the Brain and Intelligence, Jerison (1973) includes a figure showing the relationship between brain weight and body size among primates. The point is that in hominid species, brain size has developed out of proportion to body size. Jerison (1973, p. 405) has no doubt about the origin of these developments:

> "We need not look far for special selection pressures toward the major enlargements of the brain within the genus Homo. The socialized life of a predacious primate is so obviously benefited by linguistic skills, and language

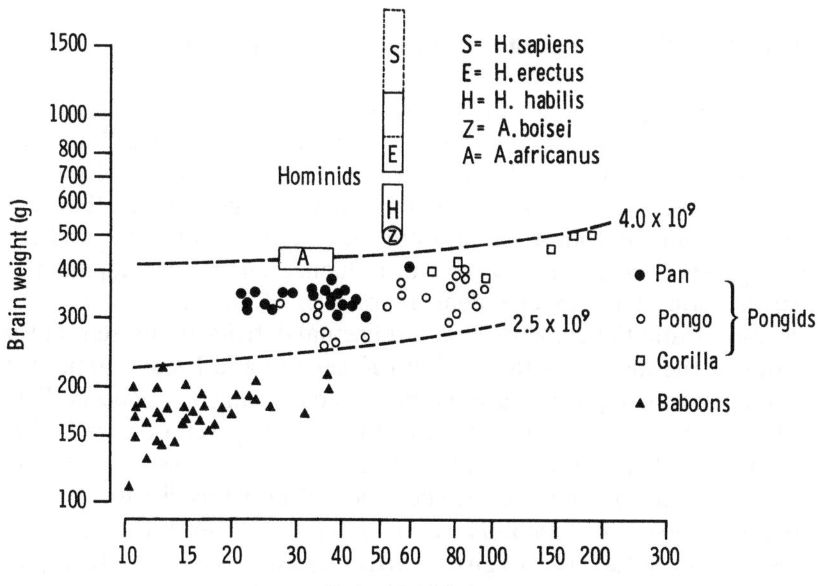

Fig. 4. Relationship between brain weight and body weight in the apes and hominid species. Dotted lines indicate the expected relationship between brain weight and body weight in species with a similar degree of encephalisation. A, australopithecus; H, homo. (Adapted from Jerison 1973)

is so manifestly the peculiarly human development, that change in the brain to permit that advantageous supplement to perception and communication would have had obvious selective advantages throughout the period of hominid evolution."

The implication of such a view is that, with the advent of language, a new dimension of variation upon which selection could act was introduced. Since the work of Dax (1835) and Broca (1861) it has been apparent that language and lateralisation of the brain are closely connected. It seems reasonable to assume that the genetic change that allowed a degree of independence in the development of the two sides of the brain also provided the substrate for the development of language. Levy (1977) has spelt out the benefits of having the two hemispheres perform different functions:

"Bilateralization of function arose in response to specifically human pressures, occurring in a socially organized species, the members of which were mutually interdependent ....."

The significance of the cerebral dominance gene then is that it permitted new evolutionary developments, and the significance of psychosis may be that it reflects the disadvantageous spin-off of this essentially human evolutionary development.

Intellectual Deficits Preceding and Succeeding Onset

In schizophrenia there are reports of intellectual deficits ante-dating onset. For example, school performance (Pollack et al. 1966) and intelligence (Lane and Albee 1964; Offord and Cross 1971) have been found to be lower in subjects who later developed schizophrenic illnesses than in their siblings. The fact that such deficits long precede the onset of, and are unrelated in form to, psychotic symptoms suggests that the disease has an influence on the development of intellectual function that is independent of episodes of illness.

There is also substantial evidence that intellectual deficits are present in the chronic forms of the disease, although both Kraepelin and Bleuler appear to have believed that true impairments of memory and learning do not occur. Thus Bleuler (1950, p. 55) wrote: "In contrast to the organic psychoses, we find in schizophrenia ..... that sensation, memory and consciousness ..... are not directly disturbed" and "memory as such does not suffer in this disease" (p. 59). Kraepelin (1919) asserted: "memory is comparatively little disordered. The patients are able when they like to give a correct detailed account of their past life, and often know accurately to a day how long they have been in the institution" (p. 18).

These views are challenged by the phenomenon of "age disorientation," aspects of which have been noted in chronic schizophrenia by a number of workers. In a survey of inpatients with schizophrenic illnesses of long standing,

Crow and Mitchell (1975) were impressed by the frequency with which such subjects believed themselves to be an age widely different from their true age. In a sample of 237 patients in the wards of four mental hospitals in Scotland, they found that approximately 25% believed themselves to be 5 or more years younger than they really were. Twelve per cent believed themselves to be within 5 years of their age on admission, although on average they were 28 years older than this. Similar findings were obtained in other surveys (Stevens et al. 1978; Smith and Oswald 1976). Age disorientation cannot be attributed to past physical treatments. Patients with age disorientation are no more likely than patients in the same institution with a diagnosis of schizophrenia, but without such cognitive deficits, to have received electroconvulsive or insulin coma therapy, a combination of these treatments, or leucotomy (Crow 1986b). Nor are they more likely to have received neuroleptic or anticholinergic medication (Buhrich et al. 1988).

Age disorientation is an aspect of a more general loss of intellectual capacity. Crow and Stevens (1978) found patients with age disorientation much less likely than those without age disorientation to be able to give correct answers to simple questions about dates and the passage of time (e.g., the present year, and the duration of their hospital stay). The errors that individual patients made in their answers to these questions were consistent with their concept of their own age. Thus age disorientation is part of a constellation of deficits of temporal orientation; for these patients, "time stands still."

Liddle and Crow (1984) assessed whether temporal disorientation is associated with general deficits of intellectual function and found, on a battery of tests of cognitive function, that age-disoriented patients were more impaired. This was true of tests of orientation and general knowledge as well as tests of new learning, problem solving and tests of previous learning and dysphasia. Thus age disorientation is but part of a global impairment of intellectual function.

It can be concluded that, contrary to the views of Kraepelin and Bleuler, intellectual impairments are sometimes seen in patients with schizophrenia, and may be profound. In some patients they precede the onset of the psychosis, apparently by many years. In other (perhaps sometimes the same) patients they progress (Buhrich et al. 1988; Garside 1969; Rappaport and Webb 1950), although whether such progression occurs in episodes of illness is unclear.

Intelligence and the Psychotic Continuum

When a broad view of psychosis is taken, there are indications that predisposition to disease is associated with benefits as well as disability. Thus Mason (1956) examined the IQ performances of inductees to the US army who subsequently required psychiatric care and compared them to the general population of inductees (Table 4).

Patients who subsequently were invalided out with a schizophrenic illness had done less well on IQ tests at induction than the average inductee, as predicted from the findings of Lane and Albee (1964). The small group of subjects who

Table 4. Comparison to control inductees to the US Army[a]

	n	P	
All patients	510	0.14	
All schizophrenic patients	368	0.02	Low scores
Manic-depressive patients	25	0.001	High scores

[a]From Mason (1956)
n: Sample size; P: probability

suffered from an affective illness had performed significantly better than their fellows at induction. Thus predisposition to affective illness is associated with an intellectual advantage at a time not far removed from onset of illness.

Further evidence of advantage in relation to the psychosis gene was identified in a study by Noreik and Ødegaard (1966) of the occupational levels of individuals admitted to psychiatric hospitals in Norway.

Table 5. Age standardised first admissions for Norway, 1926–1950[a]

	All diagnoses	Schizophrenia	Manic-depressive illness
General population	690	357	73
Professional service	704	305	151

[a]From Noreik and Ødegaard (1966)

The entire group of those ever admitted to a psychiatric hospital were as likely to have held a job in professional service (teaching, liberal professions, higher administrative posts) as the general population, while those who were diagnosed as suffering from schizophrenia were less likely to have done so. By contrast, those who were admitted with a diagnosis of manic-depressive illness were twice as likely to have held such a post. Such findings suggest that the affective form of the psychosis gene at least, has benefits in present-day society.

Such benefits are carried by the individual. More generally Karlsson (1984) has indicated that the presence of the gene in the family may be associated with social achievement. First-degree relatives of those admitted to the Reykjavik mental hospital were significantly more likely than the general population to have been listed in the Icelandic Who's Who, to have written a book and to have graduated from university (Table 6). Apparently they were also more likely to have been members of the Icelandic parliament.

Thus it may be that while possession of an allele of the psychosis gene that itself predisposes to affective psychosis confers an increased probability of success, in the case of variants of the gene that carry the risk of schizophrenia, such benefits are more apparent in first degree relatives than in the affected individual.

Table 6. Percentage rates of membership in three different classes of achievement, comparing first degree relatives of patients with psychosis with the general population, both born 1851–1940[a]

Measure	First-degree relatives (sample 9379)		General population (sample 171,000)	
	N	%	N	%
Listing in Who's Who	391	4.2	5813	3.4
Book authorship	184	2.0	2269	1.3
College graduation	234	2.5	3485	2.0

[a]From Karlsson (1984)

Other studies have indicated that, among individuals of exceptional achievement, rates of psychotic illness are at least as common as in the general population. In an earlier work Karlsson (1978) examined lists of creative individuals as indicated by their inclusion in biographical compendia, and identified those whom be thought had suffered from a psychotic illness (Table 7). The diagnostic criteria are wide and their application necessarily uncertain, but the apparent extent of psychiatric abnormality and its consistency across fields of endeavour is surprising.

In the better circumscribed field of the affective disorders, Jamison (1990) has illustrated the frequency with which such illnesses are seen in individuals of high

Table 7. Rate of recorded psychosis in creative persons listed in biographical compendia[a]

	Total listed	History of mental illness	%	Individuals who may have suffered from psychotic illness
Novelists	20	6	30	Swift, Scott, Hugo, Dostoyevsky, Tolstoy, Maupassant
Poets	20	6		Pope, Burns, Coleridge, Byron, Shelley, Poe
Painters	20	7	35	Raphael, Da Vinci, Rembrandt, El Greco, Turner, Goya, van Gogh
Composers	20	7	35	Handel, Schubert, Mendelsohn, Chopin, Schumann, Tchaikovsky, Rimsky-Korsakov
Mathematicians	32	8	25	Descartes, Pascal, Newton, Legrange, Jacobi, Hamilton, Weierstrass, Cantor
Philosophers	20	8	40	Plato, Descartes, Kant, Schopenhauer, Emerson, Spencer, Nietzsche, James

[a]Adapted from Karlsson (1978)

Table 8. Poets suffering from affective illness listed in the New Oxford Book of American Verse (total listed 36)[a]

Poet	Pulitzer Prize	Treated for major depressive illness	Treated for mania	Commited suicide
Hart Crane		+	+	+
Theodore Roethke	+	+	+	
Delmore Schwartz		+	+	
John Berryman	+	+	+	+
Randall Jarrell		+	+	+
Robert Lowell	+	+	+	
Anne Sexton	+	+	+	+
Sylvia Plath	+	+	?	+

[a]Adapted from Jamison (1990)

attainment in a variety of fields. For example, among the 36 poets included in the New Oxford Book of American Verse (Table 8), as many as eight were treated for major depressive illness, at least seven for mania, and no fewer than five committed suicide.

Cerebral Asymmetry and Intelligence

If cerebral asymmetry is the key to the recent evolutionary development of the brain that underlies human intelligence, and if as suggested by the above arguments significant variation in the underlying genetic mechanisms persists and accounts for differences in individual capacity, a relationship between the outward manifestation of such asymmetry and intelligence might be expected. Recently Annett and Manning (1989) have reported that those children who are strongly biassed toward right-handedness and who, according to Annett's theory, are likely to include among their number a high proportion of individuals homozygous for the right shift factor, are at a disadvantage relative to other children with respect to general intelligence. Thus the right shift factor and intelligence are likely to be closely related.

Conclusions

Rather than supporting the Kraepelinian notion that manic-depressive psychosis and schizophrenia are separate diseases, a number of recent studies suggest that such illnesses form part of a continuum of psychosis that extends from unipolar depression through bipolar manic-depressive illness and schizoaffective disorder to schizophrenia, with increasing severities of defect state. Although within families there is a tendency for the form of the illness in a relative to resemble that in the proband, there are also significant variations. No clear demarcation between types of psychotic illness can be made on genetic grounds. Such a pattern of illness is compatible with a high degree of variation at a single genetic locus.

A locus in the pseudoautosomal (exchange) region of the sex chromosomes is suggested to account for:

1. occasional apparent sex linkage (e.g., in the affective disorders) occurring on a background of an autosomal pattern of transmission;
2. the excess of sex chromosome aneuploidies observed in populations of patients with psychosis; and
3. concordance by sex within families in patients with both schizophrenic and affective psychoses.

A clue to the nature of the gene comes from studies in schizophrenia that show that the morphological changes in the brain are asymmetrical. One post-mortem study in particular demonstrates that such asymmetry (e.g., with respect to the temporal horn) is not present in a disease (Alzheimer-type dementia) that also causes temporal horn enlargement. Such findings suggest that there is a close relationship between the determinants of cerebral asymmetry (the right shift factor or cerebral dominance gene) and the genetic mechanisms underlying psychotic illness. Neuropsychological findings in the sex chromosome aneuploidies (XO, XXY and XXX syndromes) are consistent with a pseudoautosomal location for a cerebral asymmetry determinant.

The persistent and relatively uniform high prevalence of psychotic illness in populations across the world indicates that the underlying genetic mechanisms are likely to be of particular significance for the human condition. Speech and the associated increase in capacity for communication could have depended upon a genetic change which allowed a degree of independence in the development of the two cerebral hemispheres. Such a change may well have generated a dimension of variation upon which selection could act to increase the functional complexity of the brain. Such variation might account for differences between individuals manifest as intelligence.

That the genetic mechanisms underlying psychosis are closely related to intellectual capacity is suggested by the following observations.

1. intellectual deficits sometimes precede and sometimes follow the onset of a schizophrenic illness;
2. onset of affective illness is sometimes preceded by enhanced intellectual function;
3. predisposition to affective illness is associated with a generally high level of lifetime occupational attainment;
4. predisposition to psychosis in general is associated with above average levels of attainment in first-degree relatives; and
5. rates of affective and psychotic illness above population expectation are seen in exceptionally creative individuals.

If the genetic mechanisms underlying cerebral asymmetry are associated both with individual variations in intelligence and with risks of psychosis, it is apparent that the identification of the relevant genes will pose a number of ethical dilemmas. Understanding these mechanisms may help to elucidate the psychotic disease process and perhaps allow its course to be modified, but to use such knowledge as the basis for reproductive decisions is unlikely to be simple or free from controversy.

References

Angst J, Scharfetter C (1990) Schizoaffective Psychosen. Ein nosologisches Ärgernis. In: Lungershausen E, Kaschka WP, Witkowski RJ (eds) Affektive Psychosen. Schattauer, Stuttgart

Angst J, Scharfetter C, Stassen HH (1983) Classification of schizo-affective patients by multidimensional scaling and cluster analysis. Psychiatr Clin (Basel) 16: 254–264

Annett M (1985) Left, right, hand and brain: the right shift theory. Erlbaum, London

Annett M, Manning M (1989) The disadvantages of dextrality for intelligence. Br J Psychol 80: 213–226

Bleuler E (1950) Dementia praecox or the group of schizophrenias. International University Press, New York

Broca P (1861) Perte de la parole. Ramollisement chronique et déstruction partielle du lobe antérieur gauche du cerveau. Bull Soc Anthropol 2: 219–235

Brown R, Colter N, Corsellis JAN, Crow TJ, Frith CD, Jagoe R, Johnstone EC, Marsh L (1986) Post-mortem evidence of structural brain changes in schizophrenia. Arch Gen Psychiatr 43: 36–42

Buhrich N, Crow TJ, Johnstone EC, Owens DGC (1988) Age disorientation in chronic schizophrenia is not associated with pre-morbid intellectual impairment or past physical treatment. Br J Psychiat 152: 466–469

Burgoyne PS (1986) Mammalian X and Y crossover. Nature 319: 258–259

Collinge J, Boccio A, DeLisi LE, Johnstone EC, Lofthouse R, Owen F, Poulter M, Risby D, Shah T, Crow TJ (1989) Evidence for a pseudoautosomal locus for schizophrenia: a sibling pair analysis. Cytogenet Cell Genet 51: 978

Cooke HJ, Brown WRA, Rappold GA (1985) Hypervariable telomeric sequences from the human sex chromosomes are pseudoautosomal. Nature 317: 687–692

Crichton-Browne J (1879) On the weight of the brain and its component parts in the insane. Brain 2: 42–67

Crow TJ (1986a) The continuum of psychosis and its implication for the structure of the gene. Br J Psychiat 149: 419–429

Crow TJ (1986b) Temporal disorientation in chronic schizophrenia: the implications of an "organic" psychological impairment for the concept of "functional" psychosis. In: Kerr A, Snaith P (eds) Contemporary issues in schizophrenia. Gaskell, London, pp 168–174

Crow TJ (1987) Pseudoautosomal locus for psychosis? Lancet 2: 1532

Crow TJ (1988) Sex chromosomes and psychosis: the case for a pseudoautosomal locus. Br J Psychiatry 153: 675–683

Crow TJ (1989) Pseudoautosomal locus for the cerebral dominance gene. Lancet 2: 339–340

Crow TJ, Mitchell WS (1975) Subjective age in chronic schizophrenia: evidence for a sub-group of patients with defective learning capacity? Br J Psychiatry 126: 360–363

Crow TJ, Stevens M (1978) Age disorientation in chronic schizophrenia: the nature of the cognitive deficit. Br J Psychiatry 133: 137–142

Crow TJ, Ball J, Bloom SR, Brown R, Bruton CJ, Colter N, Frith CD, Johnstone EC, Owens DGC, Roberts GW (1989a) Schizophrenia as an anomaly of the development of cerebral asymmetry. Arch Gen Psychiatry 6: 1145–1150

Crow TJ, Colter N, Frith CD, Johnstone EC, Owens DGC (1989b) Developmental arrest of cerebral asymmetries in early onset schizophrenia. Psychiatry Res 29: 247–253

Crow TJ, DeLisi LE, Johnstone EC (1989c) Concordance by sex in sibling pairs with schizophrenia is paternally inherited: evidence for a pseudoautosomal locus. Br J Psychiatry 155: 92–97

Crow TJ, DeLisi LE, Johnstone EX (1990) In reply..... a locus closer to the telomere? Br J Psychiatry 156: 416–420
Curtis D, Gurling HDM (1990) Unsound methodology in investigating a pseudoautosomal locus in schizophrenia. Br J Psychiatry 156: 415-416
Dax M (1865) Lésions de la moitié gauche de l'éncéphale coincident avec l'oublie des signes de la pensée. Gaz Hebdom Med Chir 11: 259–260
Egeland JA, Gerhard DS, Pauls DL, Sussex JN, Kidd KK, Allen CR, Hostetter AM, Housman DE (1987) Bipolar affective disorders linked to DNA markers on chromosome 11. Nature 325: 783–787
Ferguson-Smith MA (1965) Karyotype-phenotype correlations in gonadal dysgenesis and their bearing on the pathogenesis of malformations. J Med Genet 2: 142–155
Forssman H (1970) The mental implications of sex chromosome aberrations. Br J Psychiatry 117: 353–363
Garside RF (1969) The relationship between schizophrenia and intelligence. PhD thesis, University of Newcastle
Gershon ES, Hamovit, J, Guroff JJ, Dibble E, Leckmann JF, Sceery W, Targum SD, Nurnberger JI, Goldin LR, Bunney WE (1982) A family study of schizo-affective, bipolar I, bipolar II, unipolar and normal control patients. Arch Gen Psychiatry 39: 1157–1167
Gershon ES, DeLisi LE, Hamovit J, Nurnberger JI, Maxwell ME, Schreiber J, Dauphinais D, Dingman CW, Guroff JJ (1988) A controlled family study of chronic psychoses: schizophrenia and schizo-affective disorder. Arch Gen Psychiatry 45: 328–336
Geschwind N, Levitsky W (1968) Left-right asymmetry in temporal speech region. Science 161: 186–187
Goetzl U, Green R, Whybrow P, Jackson R (1974) X linkage revisited; a further study of manic-depressive illness. Arch Gen Psychiatry 31: 665–672
Goodfellow PJ, Darling SM, Thomas NS, Goodfellow PN (1986) A pseudoautosomal gene in man. Science 234: 740–743
Haug JO (1962) Pneumoencephalographic studies in mental disease. Acta Psychiatr Scand 38: 1–104
Huber G (1957) Pneumoencephalographische und psychopathologische Bilder bei endogen Psychosen. Springer, Berlin Göttingen Heidelberg
Jamison KR (1990) Manic-depressive illness and accomplishment: creativity, leadership, and social class. In: Goodwin FK, Jamison KR (eds) Manic-depressive illness. Oxford University Press, Oxford, pp 332–372
Jerison HJ (1973) Evolution of the brain and intelligence. Academic, New York
Johnstone EC, Crow TJ, Frith CD, Husband J, Kreel L (1976) Cerebral ventricular size and cognitive impairment in chronic schizophrenia. Lancet 2: 924–926
Johnstone EC, Owens DGC, Crow TJ, Frith CD, Alexandropolous K, Bydder G, Colter N (1989) Temporal lobe structure as determined by nuclear magnetic resonance in schizophrenia and bipolar affective disorder. J Neurol Neurosurg Psychiatry 52: 736–741
Karlsson JL (1978) Inheritance of creative intelligence. Nelson-Hall, Chicago
Karlsson JL (1984) Creative intelligence in relatives of mental patients. Hereditas 100: 83–86
Kelsoe JR, Ginns EI, Egeland JA, Gerhard DS, Goldstein AM, Bale SJ, Pauls DL, Long RT, Kidd KK, Conte G, Housman DE, Paul SM (1989) Re-evaluation of the linkage relationship between chromosome 11p loci and the gene for bipolar affective disorder in the Old Order Amish. Nature 342: 238–243
Kendell RE (1988) Other functional psychoses. In: Kendell RE, Zealley AK (eds) Companion to psychiatric studies, 4th edn. Livingstone, Edinburgh, pp 363–377
Kendell RE, Gourlay J (1970) The clinical distinction between the affective psychoses and schizophrenia. Br J Psychiatry 117: 261–266
Kendler KS, Hays P (1983) Schizophrenia subdivided by the family history of affective disorder. Arch Gen Psychiatry 40: 951–955
Kendler KS, Gruenberg AM, Tsuang MT (1986) A DSM-III family study of the nonschizophrenic psychotic disorders. Am J Psychiatry 143: 1098–1105
Koller PC, Darlington CD (1934) The genetical and mechanical properties of the sex chromosomes I Rattus Norvegicus. J Genet 29: 159–173

Kraepelin E (1919) Dementia praecox and paraphrenia. Krieger, New York
Kraepelin E (1920) Die Erscheinungsformen des Irreseins. Gesamte Neurol Psychiatr 62: 1–29
Lane EA, Albee GW (1964) Early childhood intellectual differences between schizophrenic adults and their siblings. J Abnorm Soc Psychol 68: 193–195
Leader (1987) A continuum of psychosis? Lancet 2: 889–890
Levy J (1977) The mammalian brain and the adaptive advantage of cerebral asymmetry. Ann NY Acad Sci 299: 264–272
Liddle P, Crow TJ (1984) Age disorientation in chronic schizophrenia is associated with global intellectual impairment. Br J Psychiatry 144: 193–199
Luys MJ (1881) Contribution à l'étude d'une statistique sur le poid des hémispheres cérébraux à l'état normal et à l'état pathologique. Encephale 1: 644–646
Maier W, Hallmayer J, Heun R (1989) A controlled family study in affective and schizophrenic disorders: support for continuum models and spectrum concepts. First World Congress of Psychiatric Genetics, 1: 35.
Mason CF (1956) Pre-illness intelligence of mental hospital patients. J Consult Psychol 20: 297–300
McManus IC (1985) Handedness, language dominance and aphasia: a genetic model. Psychol Med [Suppl] 8: 1–40
Mendlewicz J, Rainer JD (1974) Morbidity risk and genetic transmission in manic depressive illness. Am J Hum Genet 26: 692–701
Mott FW (1910) Hereditary aspects of nervous and mental disease. Br Med J 2: 1013–1020
Netley CT (1986) Summary overview of behavioural development in individuals with neonatally identified X and Y aneuploidy. Birth Defects 22: 293–306
Netley CT, Rovet J (1982) Atypical hemispheric lateralization in Turner syndrome subjects. Cortex 18: 377–384
Netley CT, Rovet J (1987) Relations between a dermatoglyphic measure, hemispheric specialization, and intellectual abilities in 47, XXY males. Brain Cogn 6: 153–160
Noreik K, Ødegaard O (1966) Psychoses in Norwegians with a background of higher education. Br J Psychiatry 112: 43–55
Ødegaard O (1972) The multifactorial inheritance of predisposition to schizophrenia. In: Kaplan AR (ed) Genetic factors in "schizophrenia". Thomas, Springfield, pp 256–275
Offord DR, Cross LA (1971) Adult schizophrenia with scholastic failure or low IQ in childhood. Arch Gen Psychiatry 24: 431–436
Penrose LS (1942) Auxiliary genes for determining sex as contributory causes of mental illness. J Ment Sci 88: 308–316
Pollack M, Woerner MG, Goodman W, Greenberg IM (1966) Childhood developmental patterns of hospitalized adult schizophrenic and non-schizophrenic patients and their siblings. Am J Orthopsychiatry Psychiatry 36: 510–517
Rappaport SR, Webb WB (1950) An attempt to study intellectual deterioration by premorbid testing. J Consult Psychol 15: 95–98
Reveley MA, Reveley AM, Baldy R (1987) Left hemisphere hypodensity in discordant schizophrenic twins: a controlled study. Arch Gen Psychiatry 44: 625–632
Rice J, Reich T, Andreasen N, Endicott J, van Eerdewegh M, Fishman R, Hirschfeld RMA, Klerman GL (1987) The familial transmission of bipolar illness. Arch Gen Psychiatry 44: 441–447
Rosenthal D (1962) Familial concordance by sex with respect to schizophrenia. Psychol Bull 59: 401–421
Smith JM, Oswald WT (1976) Subjective age in chronic schizophrenia. Br J Psychiatry 128: 100
Southard EE (1915) On the topographic distribution of cortex lesions and anomalies in dementia praecox with some account of their functional significance. Am J Insanity 71: 603–671
Stassen HH, Scharfetter C, Winokur G, Angst J (1988) Familial syndrome patterns in schizophrenia, schizoaffective disorder, mania, and depression. Eur Arch Psychiatry Neurol Sci 237: 115–123

Stevens M, Crow TJ, Bowman MJ, Coles EC (1978) Age disorientation in schizophrenia: a constant prevalence of 25 % in a chronic mental hospital population? Br J Psychiatry 133: 130–136
Weissenbach J, Levilliers J, Petit C, Rouyer F, Simmler M-C (1978) Normal and abnormal interchanges between the human X and Y chromosomes. Development [Suppl] 101: 67–74
Winokur G, Clayton PJ, Reich T (1969) Manic-depressive illness. Mosby, St Louis

Molecular Genetic Studies: The Task of Interpretation for Psychiatry

M. Baron

Introduction

That genetic factors operate in some psychiatric disorders, especially schizophrenia and major affective illness, is generally not disputed. These disorders tend to run in families; the risk to relatives of affected individuals is several times greater than the base rate in the population. Twin and adoption studies suggest that the familial aggregation is due in part to heredity. However, little is known about the underlying genetic mechanism.

Until recently, attempts to elucidate genetic susceptibility to psychiatric illness had focused primarily on statistical models that test the fit of specific genetic hypotheses to family data, and on potential biological risk factors. These efforts by and large have been frustrated. Various mathematical models invoking a single major gene, a small number of interacting genes, multifactorial-polygenic transmission, and major gene effects with multifactorial background have failed to discern with certainty mode (or modes) of inheritance (Baron 1986a,1991; Gershon et al. 1989). An extensive search for biological traits that may correlate with genetic susceptibility, such as neurotransmitter enzymes and metabolites, hormones, receptor proteins, and neurophysiological and brain imaging indices, has yielded inconsistent, incomplete or otherwise ambiguous results (Baron 1986a, b, 1991; Gershon et al. 1989).

Recent abvances in molecular biology have revolutionized genetic research. Noteworthy among these developments are DNA markers, also known as restriction fragment length polymorphisms (RFLPs). These are inherited differences in base pair sequences of DNA which can be detected using radiolabeled DNA probes. RFLPs segregate in a Mendelian fashion. They span the entire human genome and can thus be used, primarily through linkage analysis, to localize genes for various diseases.

Linkage refers to the tendency of a particular marker allele to co-segregate with the disorder within a pedigree. This is likely to occur when the marker locus and the disease locus are in close proximity on the same chromosome. Thus, linkage studies aim to detect linkage and to determine the approximate distance between loci. Linkage can be measured by calculating the lod score (the logarithm of odds in favor of the linkage hypothesis). Conventionally, when the lod score exceeds 3 (corresponding to odds of 1000 to 1), linkage is considered present. The genetic distance between loci is estimated by the recombination

Srám, Bulyzhenkov, Prilipko, Christen (Eds.)
Ethical Issues of Molecular Genetics in Psychiatry

fraction, a statistical measure of crossover, or material exchange between homologous chromosomes during meiosis. The principles of linkage analysis were reviewed elsewhere (Ott 1985).

Once linkage has been established, attempts can be made to bridge the map distance between the marker locus and the disease locus. The identification of tightly linked markers and, ultimately, the cloning of the disease gene have major implications for risk prediction (genetic counselling) and for the elucidation of the disease process. The principles and potential applications of these methods, with special reference to psychiatric disorders, have been discussed elsewhere (Gurling 1985; Baron and Rainer 1988).

Numerous disease genes have already been located by these techniques, including the genes for the neuropsychiatric disorders Huntington's disease and a familial form of Alzheimer's dementia (Kidd et al. 1989). In many disorders, such as hemophilia, neurofibromatosis, some forms of muscular dystrophy, and polycystic kidney disease, diagnostic testing using molecular genetic techniques is already available. In some instances, most notably Duchenne muscular dystrophy and cystic fibrosis, the gene itself has been cloned.

The impact of the 'new genetic's on psychiatry is already apparent. Several studies have reported evidence for linkage between psychiatric illness and loci mapped to specific genomic regions. These include studies supporting linkage of bipolar affective illness to genetic markers on the X-chromosome (Baron et al. 1987; Mendlewicz et al. 1987) and chromosome 11 (Egeland et al. 1987) and of schizophrenia to chromosome 5 loci (Sherrington et al. 1988). However, these studies were followed in quick succession by negative linkage reports. The inconsistent, and seemingly contradictory, data are somewhat bewildering. They complicate the task of interpreting the various results and charting a course for future research. In what follows, I will discuss some of the major methodological issues which face the researcher and, ultimately, the clinician in this area of investigation. I will then interpret the recent genetic findings in the context of these potential uncertainties. A more detailed treatment of these issues has appeared elsewhere (Risch 1990; Baron et al. 1990a, b).

True versus False Findings

A common explanation for diverse linkage findings is etiologic heterogeneity; that is, the presence of different genetic and non-genetic determinants in etiology. However, in the face of phenotypic and genetic complexities, spurious linkage results are a distinct possibility. The task at hand, then, is to distinguish genuine from false linkage findings. The latter category encompasses two classes of results: false positive (a false claim that linkage exists) and false negative (rejection of linkage when true linkage exists).

False positive linkage can arise under the following circumstances:

1. Systematic error in the assignment of genotypes and/or disease phenotype; for example, non-blindness of disease versus marker status and vice versa.

2. Selection of a favored set of genetic parameters and/or disease definition from multiple analyses in an attempt to maximize the lod score (and thereby inflating the evidence for linkage).
   This scenario may happen in complex disorders with unknown genetic parameters and a variable phenotype. For example, because lod score analysis requires specification of penetrance and allele frequencies, a common practice is to employ several sets of these parameters. Similarly, what constitutes the phenotypic manifestation of the disease genotype is far from certain. The so-called familial 'spectrum' of psychiatric disorders encompasses a large variety of conditions that aggregate in the families of affected individuals. In schizophrenia these range from the chronic schizophrenic type, which presumably represents the severe end of the spectrum, to milder clinical variants, such as schizotypal and paranoid personalities (Baron 1986a). In affective disorders, the spectrum ranges from bipolar illness, arguably the severe phenotype, to milder conditions which include depressive disorders, cyclothymia and hypomania (Gershon et al. 1989; Baron 1991).
3. Because a genomic search for disease-related genes usually involves multiple genetic markers, a positive linkage finding may occur by chance.

False negative linkage may be a more common source of error. Given the likely heterogeneity, if a particular disease form is rare, linkage can be easily missed unless a large series of pedigrees is studied. True linkage may be concealed due to other complications, such as misspecified genetic parameters and, more importantly, diagnostic uncertainties leading to high rate of phenocopies, and heterogeneity within the pedigree for other reasons, such as cohort effect and bilineal transmission (e.g., assortative mating). These potential complications are commonplace in psychiatric genetics. As noted earlier, estimates of genetic parameters are bound to be crude due to the complex mode of transmission. Also, because some of the phenotypes are common in the population (e.g., depression), phenocopies are certain to occur. Uncertainties concerning diagnostic reliability and stability, especially in the case of the milder psychiatric disorders, constitute another potential source of phenocopies. Cohort effect, which may be partly due to nongenetic factors, has been observed in affective disorders, and assortative mating or other forms of bilineal transmission exist both in schizophrenia and in affective illness.

Finally, the molecular genetic approach is not foolproof. Typing errors do occur due to complex allelic patterns, sample mix-ups or other technical difficulties.

Safeguards Against Spurious Linkage Findings

Measures can be taken to minimize the occurrence of false linkage results. First, diagnostic determination and marker typing should be carried out independently.

Second, decisions concerning disease classification and genetic parameters should be made prior to the analysis. The number of classifications should be limited and constrained to a 'realistic' range; for example, disorders that are not known to aggregate in families need not be included in the disease phenotype.

Third, more credence should be accorded to a conservative approach whereby the disease is defined by narrow criteria; this should enhance the certainty with which the clinical diagnosis is made. Broader criteria can be used subsequently, preferably with allowances for diagnostic uncertainty and phenocopies. For example, estimates of the stability of diagnosis based on follow-up data can be incorporated in the analysis (Baron et al. 1990b). To avert misclassifications among the presumed unaffected, a 'penetrance free' model can be used in which only strictly defined affected cases are considered in the analysis. Diagnostic assessments should be carried out by experienced clinicians using standardized research procedures and subject to frequent checks on inter-rater reliability.

Fourth, priority should be given to pedigrees with unilineal transmission to the exclusion of assortative mating; this requires a thorough investigation of marry-in spouses and their relatives.

Fifth, consideration of the lod score criterion should reflect statistical adjustment for the number of markers tested, especially in the case of an extensive genomic search. Procedures for establishing the reliability of laboratory results should be routine.

Sixth, replication is the ultimate 'proof' that a particular finding is genuine. However, if a particular linkage accounts for a small fraction of cases, a successful replication may require a large sample and considerable resources. Other methods can be employed to evaluate linkage results. These include:

1. extension of genetic information by adding new individuals and typing additional markers in reported pedigrees; and
2. longitudinal follow-up to assess diagnostic stability and late illness onset.

Should the reanalysis fail to sustain the original finding, the latter must have been a false positive result. If the original finding is sustained, confidence in the particular linkage reported will be increased.

Finally, there is uncertainty about the proper lod score criterion for complex disorders. The statistical method was devised for simple Mendelian disorders in which the aformentioned complications do not exist. The lod criterion, therefore, may have to be made more stringent (that is, higher than the conventional value of 3.0) to account for the uncertainties underlying the analysis.

Recent Studies: Description

Affective Disorder and Chromosome 11

In 1987, linkage was reported between bipolar affective disorder and DNA markers at the insulin and HRAS loci on the short arm of chromosome 11 (11p;

Egeland et al. 1987). The report was based on a large Old Order Amish pedigree. The maximum lod score reported by the investigators was nearly 5.0. The result was also of interest because the gene for tyrosine hydroxylase, a pivotal enzyme in biogenic amine metabolism, is located in close proximity to the putative disease locus. Several research groups have failed to replicate this finding. Moreover, two years later, on extension and renalysis of the original pedigree data, the evidence for linkage has all but vanished (Kelsoe et al. 1989). The revised outcome of the study was attributed to changed diagnoses in two individuals, as well as to a different segregation pattern in the added branches of the pedigree.

Affective Disorder and the X-Chromosome

Shortly after the original report on the Amish, linkage was reported between bipolar affective illness and the Xq28 markers color blindness and glucose-6-phosphate dehydrogenase (G6PD) deficiency in several Israeli pedigrees (Baron et al. 1987). The overall maximum lod score was 7.5; linkage appeared more pronounced in the non-Ashkenazi group of pedigrees where the lod score climbed to 9.2. Subsequently the study was extended in two ways:

1. the data were reanalyzed using narrower diagnostic criteria and allowing for phenocopies (Baron et al. 1990b). The lod score was somewhat reduced but the results overall were robust with respect to the disease phenotype; and
2. the pedigrees were typed for DXS98, a DNA marker locus in Xq27, in an attempt to better delineate the region containing the putative disease gene (Freimer et al. 1989). Linkage to DXS98 was excluded.

Another recent study of the X-linkage hypothesis yielded a lod score of 3.1 for linkage between bipolar illness and a DNA marker at the F9 locus in Xq27; the pedigrees were obtained from a Belgian population (Mendlewicz et al. 1987). The lod score, however, was critically dependent on the penetrance value; for penetrances lower than 100 % the lod score was smaller than 3.0.

Schizophrenia and Chromosome 5

Spurred by a report a chromosomal anomaly involving chromosome 5q and the localization of the glucocorticoid receptor gene to the same region, a linkage study was carried out in several Icelandic and British pedigrees using two marker loci, D5S76 and D5S39, in the 5q11 region (Sherrington et al. 1988). The highest lod score obtained was 6.5 when the disease phenotype was broadened to include so-called 'fringe' disorders which ordinarily are not included in the schizophrenia spectrum. When the analysis was restricted to schizophrenia, the lod score was reduced to 3.2, a substantially smaller value. All subsequent studies carried out in other populations did not detect linkage to 5q markers.

Recent Studies: Interpretation

How can we interpret these studies in light of the aformentioned methodological considerations? The virtual disappearance of linkage in the Amish pedigree casts doubt on the original finding. It suggests that the initial lod score might have been too small to indicate true linkage. The non-replication reports which followed are therefore not surprising.

The evidence for linkage of bipolar illness to Xq28 markers appears stronger, with a lod score much higher than the conventional lod threshold. That it remained statistically significant with a stringent disease definition and other allowances for diagnostic uncertainties should increase confidence in the reported linkage. Pedigrees unlinked to Xq28 markers notwithstanding, earlier reports of linkage to these markers, though comparatively lacking in methodological rigor (for review, see Baron et al. 1981; Risch and Baron 1982) may therefore be viewed as consistent with the later report (Baron et al. 1987). The situation is somewhat different when the reported linkage to the F9 locus is considered (Mendlewicz et al. 1987). Here, as noted earlier, the lod score was of borderline statistical significance, the location of the purported disease gene was too distant to be compatible with the stronger evidence for an Xq28 locus, and linkage to F9 was ruled out in the pedigrees in which linkage to Xq28 appeared to be present (Freimer et al. 1989). These findings raise the possibility that the F9 linkage was a chance finding.

As for the proposed linkage between schizophrenia and chromosome 5 loci, the lod score for the conventional disease phenotype was only slightly over 3.0. All of the subsequent studies led to non-replication.

Of the three reported linkages, the linkage data on bipolar affective illness and Xq28 markers seem closest to suggest a genuine finding. This, however, requires further scrutiny according to the guidelines explicated earlier.

Implications

When a particular linkage is established, another level of interpretation is in order.

1. Ordinarily, the map position of the linkage marker is some distance away from the disease locus; namely, recombinations can be observed. Owing to the probability of error in determining the segregation of alleles at the two loci, a finding of this type does not have immediate impliations for risk prediction and genetic counseling. To reduce the level of error, it will be necessary to develop flanking markers tightly linked to the disease locus. Ideally, the gene itself should be cloned.
2. Another important goal is to determine the extent of etiologic heterogeneity. Linkage to separate loci may be required to substantiate genetic (non-allelic) heterogeneity. To evaluate with some precision the prevalence of the various genetic forms in the population, markers in linkage disequilibrium with the disease gene will be needed, preferably a probe specific for the gene itself.

This should facilitate population screening without resorting to family resources.

3. Although most of the current research in this field concerns the search for major genes, it must be recognized that other forms of genetic transmission, in particular interaction between loci, may exist. This type of genetic causation may be especially relevant to disorders that are maintained in the population in spite of strong selective disadvantage (i.e., reduced reproduction). Schizophrenia is one such example. When the number of loci involved is small, they can be detected by current methods. However, new molecular biology techniques in conjunction with quantitative genetic methods will be required to unravel the contribution of multiple loci, each of a small effect, to disease susceptibility. Even when a major gene is identified by linkage analysis, its mode of action may be influenced by minor locus effects generated by other genes. The elucidation of the various genetic contributions will be required to discern the overall genetic pattern.
4. The ultimate goal in this line of investigation is to isolate and clone disease-related genes. In the absence of suitable candidate genes, and barring a fortuitous cytogenetic abnormality that could expedite this effort, the attainment of this goal will require large amounts of genetic information (i.e., family material, DNA sequences that span the chromosomal region of interest) and will likely be labor intensive.
5. The role of nongenetic factors (e.g., psychological stress, infectious or toxic agents) should not be ignored. That such factors may operate alongside the genetic disposition can be surmised both from data on discordant identical twins and from the presence of carriers (unaffected individuals who possess the allele predisposing to the illness). Nongenetic factors can interact with genetic susceptibility either to prevent or trigger the expression of psychopathology. Linkage data, or better yet the isolation of a candidate gene, can offer a paradigm for unraveling such interactions. For example, affected and unaffected relatives who are identical at the marker locus can be studied for the presence or absence of particular environmental influences. To be sure, genetic factors other than those identified through the particular linked locus can fulfill a similar role. The elucidation of the various influences that may operate in conjunction with the genetic susceptibility will be instrumental both for risk prediction and for sorting out incomplete penetrance. Finally, some familial forms of the illness may be purely, or chiefly environmental. This possibility deserves recognition in our search for causative factors in etiology.

Summary

Recent advances in molecular biology have revolutionized medical genetics. Genetic linkage studies coupled with recombinant DNA technology may play a key role in delineating genes underlying some forms of mental illness. The potential embedded in the 'new genetics' must, however, be viewed in the context of methodological issues such as etiologic heterogeneity, phenotypic diversity, complex modes of inheritance and diagnostic uncertainties. The importance of these issues for the design and interpretation of molecular genetic studies in psychiatric disorders is discussed.

Acknowledgments. Supported by NIMH Research Scientist Development Award MH00176 and by NIMH grants MH42535 and MH43979.

References

Baron M (1986a) Genetics of schizophrenia I. Familial patterns and mode of inheritance. Biol psychiatry 21: 1051–1066

Baron M (1986b) Genetics of schizophrenia II. Vulnerability traits and gene markers. Biol Psychiatry 21: 1189–1211

Baron M (1991) Genetics of manic depressive illness: current status and evolving concepts. In: Genes, brain and behavior. McHugh PR (ed) Raven, New York, pp 153–164

Baron M, Rainer JD (1988) Molecular genetics and human disease: implications for modern psychiatric research and practice. Br J Psychiatry 152: 741–753

Baron M, Rainer JD, Risch N (1981) X-linkage in bipolar affective illness: perspectives on genetic heterogeneity, pedigree analysis and the X-chromosome map. J Affect Dis 3: 141–157

Baron M, Risch N, Hamburger R, et al. (1987) Genetic linkage between X-chromosome markers and bipolar affective illness. Nature 326: 289–292

Baron M, Endicott J, Ott J (1990a) Genetic linkage in mental illness: limitations and propects. Br J Psychiatry 157: 645–655

Baron M, Hamburger R, Sandkuyl LA, et al. (1990b) The impact of phenotypic variation on linkage analysis: application to X-linkage in manic-depressive illness. Acta Psychatry Scand 82: 196–203.

Egeland J, Gerhand D, Pauls D, et al. (1987) Bipolar affective disorders linked to DNA markers on chromosome 11. Nature 325: 783–787

Freimer N, Baron M, Risch N, et al. (1989) Multipoint linkage analysis of bipolar disorder and DNA markers in the Xq27–28 region. First World Congress on Psychiatric Genetics, Cambridge

Gershon ES, Berrettini WH, Goldin LR (1989) Mood disorders: Genetic aspects. In: Kaplan HI, Sadock BJ (eds) Comprehensive textbook of psychiatry. Baltimore, Williams and Wilkins, pp 879–887

Gurling HMD (1985) Application of molecular biology to mental illness. Analysis of genomic DNA and brain mRNA. Psychiatr Dev 3: 257–273

Kelsoe JR, Gings EI, Egeland JA, et al. (1989) Reevaluation of the linkage relationship between chromosome llp loci and the gene for bipolar affective disorder in the Old Order Amish. Nature 342: 238–243

Kidd KK, Klinger HP, Ruddk FH (ed) (1989) Human gene mapping 10 Cytogenet Cell Genet 51: 1–1147

Mendlewicz J, Simon P, Sevy S, et al. (1987) Polymorphic DNA marker on X-chromosome and manic depression. Lancet 2: 1230–1232

Ott J (1985) Analysis of human genetic linkage. Johns Hopkins University Press, Baltimore

Risch N (1990) Genetic linkage and complex diseases, with special reference to psychiatric disorders. Genet Epidemiol 7: 3–16

Risch N, Baron M (1982) X-linkage and genetic heterogeneity in bipolar related major affective illness: reanalysis of linkage data. Ann Hum Genet 46: 153–166

Sherrington R, Brynjolfsson J, Petursson H, et al. (1988) Localization of a susceptibility locus for schizophrenia on chromosome 5. Nature 336: 164–167

The Genetics of Alzheimer's Disease and the Ethical Implications for Prevention

P. McGuffin, M. Sargeant, and *G. Weppner*

Senile dementia of the Alzheimer type is probably the commonest form of dementing illness after the age of 65 years in most western populations. The classical presenile form of Alzheimer's disease is considerably less common but it is again among the most frequent causes of cognitive decline in late middle age. Since the presenile and senile forms show the same characteristic histopathology, they are now both often referred to simply as 'Alzheimer's disease' (AD). Both types tend to show familial aggregation and this seems to be particularly marked in early onset cases. Despite the consistency of reports of familiality and despite recent improvements in the understanding of the neuropathology and neurochemistry of AD, the precise aetiology remains a puzzle. This would doubtless be a disappointment to Alzheimer, who described AD and its typical neuropathological lesions in 1907, not long after his head of department, Kraepelin (1922), had effected his famous separation of the psychoses of adult life into dementia praecox and manic depressive insanity. Both men believed that they had classified conditions with a constitutional basis but only Alzheimer had clear neuropathological evidence on his side. More than 80 years on, the three conditions continue to present major challenges for genetic research and for therapeutic endeavour but, if anything, AD is the most aetiologically enigmatic and the most difficult to treat.

Our principal brief here is to discuss ethical aspects of the genetics of AD and prevention of the disorder. We will therefore begin with a short consideration of the general issues which arise in dealing with the genetics of any neuropsychiatric illness. It is impossible to proceed from the general to the specific without a knowledge of the empirical research data, and we will therefore attempt to summarise the current state of knowledge on the molecular biology and genetic epidemiology of AD before going on to attempt some explicit statement concerning ethics and the prevention of AD.

Ethical Problems and Neuropsychiatric Illness

Elsewhere in reviewing ethical issues in psychiatric genetics, we have suggested that it is best to consider the important questions under two broad headings relating first to research and second to clinical practice. We acknowledge that this is somewhat arbitrary: the two can never be cleanly separated since clinical

Srám, Bulyzhenkov, Prilipko, Christen (Eds.)
Ethical Issues of Molecular Genetics in Psychiatry

practice is always informed by research and vice versa. However, considering the issues under the two headings can sharpen our focus and facilitate description.

Research Issues

Perhaps the most fundamental ethical question facing psychiatric genetics is, to paraphrase Samuel Johnson, not so much whether it can be done well but whether it should be done at all. In a recent paper on ethics and psychiatric genetics (McGuffin and Sargeant 1989) we noted that the very suggestion that certain forms of mental illness have a genetic basis can still raise the hackles of some clinicians. Subsequent correspondence has confirmed our suspicion that in some quarters, there is a lingering animosity which is a hangover from the days of the German Third Reich. There is no doubt that early eugenicists, not just those in Germany, enthusiastically over stated their case (Roll-Hansen 1988). Furthermore, it is not improbable that since Germany in the 1920s and 1930s was the birthplace of psychiatric genetics (indeed it could be argued the birthplace of modern psychiatry), some pioneers of psychiatric genetics supported the Nazi movement. We also know that the Nazis concocted a bone-headed travesty of genetic theory to justify their barbaric treatment of the mentally ill (Meyer 1988). Although subsequently such policies were seen as not just morally repugnant but scientifically idiotic by geneticists everywhere, it is a sad fact that psychiatric genetics still suffers sometimes from a sort of guilt by association.

A less tangible but still potent source of antipathy to psychiatric genetics is the view that genetic explanations of psychological disorders are unacceptably mechanistic and demeaningly simplistic. This argument is often allied with another that claims that not only do genetic theories reduce the richness and complexity of human experience to mere molecules but also they tell us that psychiatric illness is inborn and immutable. Fortunately there is now a generally increasing awareness that nothing could be further from the truth. The recent history of the application of recombinant DNA technology in medicine (Weatherall 1985) has been one of successive triumphs. Marked advances in the understanding of the genetic basis of diseases with simple mendelian modes of transmission have taken place which offer logical preventive strategies and suggest rational bases for the development of treatments (Davies and Read 1988). Therefore, we believe that the argument that genetic research in neuropsychiatric illness offers much for the general good far outweighs any counter argument that genetic research is still open to misuse and misinterpretation.

Clinical Genetics

Broadly, the areas which are relevant to the prevention of AD are genetic counselling and predictive testing. In genetic counselling, the aim is to provide information and advice to patients, their families and to prospective spouses.

Since AD usually occurs in late adult life or in old age, it is more often the offspring of AD sufferers who seek guidance than the patients themselves. An associated problem which is sometimes presented to the genetic counsellor is of prospective adopting parents of children who have a history of mental illness in biological relatives. However, in our experience this is a rare event in association with AD. In any of these situations, the main ethical questions are: "In what form should information be given?" and "How much should the counsellor give guidance and advice as well as simply educate the counsellee?" Most clinicians now take the view that genetic counselling should be a non-directive and mainly information-imparting process. This entails an attempt to provide accurate risk figures and to dispel any mistaken beliefs that all the offspring of a parent with serious mental illness are necessarily afflicted by 'hereditary taint'?

We believe it is correct to emphasise the counsellee's personal autonomy and that a non-coercive approach should be the general rule in practice. However, it may be justifiable not just to lower anxiety by providing information and reassurance but also to raise the individual's awareness of potential problems (and hence provoke some anxiety) where this is appropriate.

Murphy and Chase (1975) have pointed out that assessment of risk in genetic counselling can be derived from three sources. These are:

1. empirical information, consisting of estimates based on actual research data;
2. modular information, which depends on a clear understanding of the mode of inheritance of the disorder; and
3. particular information which is a compilation of all the data that can be utilised in assessing the risks of a particular case.

Clearly, the counseller who wishes to achieve and maintain ethical standards needs a good scientific understanding of the source of risk figures, the current optimal approach for arriving at accurate risk figures, and the limitations as well as the possibilities of current techniques in risk prediction.

Predictive Testing

A wide range of techniques is available for prenatal diagnosis of genetic disorders, including amniocentesis, chorion biopsy, fetoscopy and ultrasound scanning, but most of these are of principal relevance to disorders and developmental anomalies which are evident in infancy (for review, see Harper 1988). For those disorders which are not likely to manifest themselves until later in life, the techniques which are most relevant involve prenatal prediction by genetic linkage. These are of particular potential benefit in neuropsychiatric conditions where, for the most part, the underlying biochemical defects are obscure and where discovery of a linked DNA polymorphism can provide a possible method of tracking the disorder. To date the only good practical model is Huntington's disease (HD), where it has now been well established in family linkage studies that there is a gene close to the tip of the short arm of chromosome 4. Since the published report of the first linked DNA polymorphism seven years ago (Gusella et al. 1983), several more close linked markers have now been discovered. As a result of this, predictive testing in pregnancy has been established and evaluated in several specialist centres (Harper and Morris 1990). Because HD does provide a potential model for future work on AD and other neuropsychiatric disorders, it is worth considering in a little detail.

The most common application is predictive testing in pregnancy, is when a healthy individual at risk of HD decides to have children and wishes to know the risk for an unborn child. The at-risk counsellee typically has an affected parent and hence, because HD is a regular autosomal dominant condition, they themselves have a 50 % risk of developing the disorder. If information is available with respect to the marker locus for both the affected and the unaffected parents, as well as the counsellee, it is then often possible by examining fetus material to say whether the unborn child has one allele or no alleles in common with both the counsellee and his or her affected parent. If the fetus possesses one such allele, then the risk of later developing HD is approximately the same as that of the counsellee (i.e., on average 50 %). If on the other hand the fetus has no such allele, then its risk is closer to zero (Quarrell et al. 1987). The exact risk depends upon the frequency of recombination between the marker locus and the HD locus, but for most currently used DNA polymorphisms this is well under 5 %.

The availability of this test provokes an obvious moral dilemma. If the fetus' risk is close to zero no difficulty arises, but if the result is adverse, the counsellee must decide whether or not to continue with the pregnancy where the child's risk of HD is approximately the same as the counsellee's own risk. Such a test is not yet available in AD but some research, as we shall discuss below, suggests that it might become available at least in a minority of families where AD segregates in a dominant-like fashion.

Perhaps an even more pointed dilemma for the clinician is that of predictive testing for at risk individuals themselves. Again HD offers a ready-made model, as a condition which results in dementia but does not usually present until adult life. Due to the discovery of linked DNA markers on chromosome 4, young adults at risk of developing HD, providing they have an informative family structure and relatives available for testing, can now be told that their probability of developing this fatal degenerative disease is either less than 5 % or more than 95 %. Given that there is no cure for HD it is a matter of debate how to provide such a predictive service and indeed whether such a service should be provided at all. Some useful guidelines have emerged as a result of early experience in predictive testing and it seems strongly advisable that testing should only be offered in centres which can precede the test with expert counselling and can provided facilities for psychological support when the outcome is adverse (Fahy 1989; Morris et al. 1989). There are compelling arguments against performing a predictive test as part of the assessment of a child who is being offered for adoption (Morris et al. 1988), and we would argue that it is a dubious practice to perform predictive testing of this type on any child.

Again, although we can look to HD as a model, the analogy is imperfect. AD is by contrast a complex condition which may be heterogeneous. Although some authors suggest that simple autosomal dominant forms exist, the overall picture is by no means clear. Therefore, before we can go further in discussing the ethics of applying either predictive testing or genetic counselling strategies in the prevention of AD, we need to consider the current status of the genetic evidence.

The Genetics of Alzheimer's Disease

It is customary when discussing the genetics of any complex disorder to commence with 'classical' studies, including investigation of familial aggregation and twins and adoption studies, if available, before going on to consider molecular biology. However, we propose to do just the reverse. Our main reason for doing this is that the molecular aspects of the genetics of AD have not only been a focus of much activity and of great interest, but also, because molecular genetics is a topical area which might offer quick returns, it demands immediate attention. After this discussion we will reflect on a more general problem of genetic research in AD, which is that there is a paucity of rigorous genetic epidemiological information. Furthermore, the information that is available has tended to be overlooked, partly, we suspect, because it is old. We will therefore consider genetic epidemiological studies second and try to suggest what these have to offer in the interpretation of recent molecular biological findings.

Molecular Genetics

Most successful attempts to pinpoint the loci for autosomally inherited disorders have been achieved with a good measure of luck. A common strategy has been to carry out a 'random' genetic linkage study with whatever polymorphisms are available in families in which the disorder is segregating. The aim then, without any prior hypothesis about where in the human genome to look, is to study the cosegregation of the disease and marker traits at a number of different loci in the hope of detecting linkage and being able to estimate recombination. For example, in HD this search was carried out in various centres for a wide variety of so-called 'classical' markers, (e.g., blood groups, HLA types, red cell enzymes, blood protein polymorphisms, etc.) with uniformly negative results (Volkers et al. 1980). Subsequently when using DNA polymorphisms, Gusella et al. (1983) had the great good fortune to chance upon linkage with their eighth marker. The availability of many DNA markers which provide a nearly complete human genetic linkage map (Donis-Keller et al. 1987) allows researchers to adopt a much more methodical approach and attempt a systematic search for disease genes throughout the entire genome. However, this is clearly likely to be a costly exercise and will not only require large samples of (preferably) multiply affected families but also require that literally hundreds of markers be studied. Not surprisingly, most investigators will attempt to find ways to take a short cut. Essentially there are three ways that a short cut can be achieved. The first is that some previous investigator, in a 'random' study, may have uncovered a hint of linkage with some marker which requires testing on a new sample. Second, there may be a clue as to the chromosomal position of a disease locus from observations of individuals with chromosomal anomalies. Third, the researcher may have some prior knowledge of the biochemistry of the condition and hence wish to focus on genes encoding for proteins which may be involved in the aberrant biochemical process. In the case of AD both the second and the third type of pointers exist.

The almost universal occurrence of cognitive decline and associated premature development of Alzheimer type changes in the brains of patients with Down's syndrome dying in their forties incriminated chromosome 21 in the pathogenesis of AD. However, one group of workers presented an alternative explanation and claimed to have found that there was also an increased risk of Down's syndrome among the relatives of index cases with AD (Heston and Mastri 1977). It was proposed that some inherited structure at a cellular level predisposed both to non-disjunction in germ cells, resulting in trisomy 21, and AD type degenerative changes in brain cells. Attempts to replicate this finding failed (Whalley et al. 1982) but, by contrast, the premature AD type decline of patients with Down's syndrome has been firmly established. The other pointer to chromosome 21 comes from studies of characteristic AD lesions. Essentially, the typical post mortem changes in the brains of patients dying with AD are of extracellular plaques containing amyloid, widespread deposition of amyloid around cerebral blood vessels, and intracellular tangles consisting of paired helical filaments (PHF). It is probable, although not absolutely certain, that PHF also consist of amyloid protein. It was therefore of great interst that, within a short period, two findings emerged which appeared to complement each other and appeared to provide an important leap forward in the understanding of the genetics of AD. One of these was a study of four large pedigrees containing multiple members affected by AD, which showed linkage between the disorder and DNA markers (known as D21sl/s11 and D21s16) which map to the long arm of chromosome 21 (St. George-Hyslop et al. 1987). The other was that the gene encoding for amyloid precursor protein (APP) was cloned and localised to an adjacent region on the long arm of chromosome 21 (Goldgaber et al. 1987; Tanzi et al. 1987). Not surprisingly, therefore, there was intense speculation that the APP gene would also be the site of the primary defect in AD. Such speculation was further encouraged by a report that three patients with 'sporadic' AD actually had an extra copy of the APP gene. In this respect they were found to resemble patients with typical trisomy 21 Down's syndrome and two patients with so-called karyotypically normal Down's syndrome (Delabar et al. 1987).

However, the excitement began to abate when it was discovered that the putative AD gene and the gene encoding for APP were probably not one and the same. Indeed they were not even tightly linked, and genetic studies and physical mapping studies confirmed that they were some distance apart (Tanzi et al. 1988; Gardiner et al. 1988). Disappointingly, several studies then failed to replicate the finding of an extra copy of the APP gene in patients with 'sporadic' AD (Podlisny et al. 1987; Warren et al. 1987). (It should be noted that in this context, the term 'sporadic' is used in a loose and, in our view, incorrect sense, to denote patients in whose families no secondary cases of AD had been detected. We will return to this point later in discussing the results of segregation analysis).

Further blows for the chromosome 21 hypothesis were delivered by two studies which failed to show linkage between AD and chromosome 21 markers (Schellenberg et al. 1988; Pericak-Vance et al. 1988). These findings were complicated by the fact that, in contrast with the earlier positive study (St. George-Hyslop et al. 1987), pedigrees which were investigated were a mixture of early onset and late-onset forms of AD. Furthermore, one of the studies (Schellenberg et al. 1988) contained pedigrees from an inbred and comparatively genetically isolated community. A subsequent study (Goate et al. 1989) which was based on non-inbred families, multiply affected with early onset AD, found supporting evidence for linkage and multipoint analysis suggested that the AD gene is closest to the D21s16 marker. This was in keeping with the earlier positive study of St. George-Hyslop et al. (1987), and interestingly the data of Pericak-Vance et al. (1988) also provided positive (albeit weakly positive) lod scores close to this marker. It is now known from physical mapping studies (Owen et al. 1990) that D21s16 is some distance centromeric of D21s/sll, the marker which has received most attention and the one with which linkage can be most convinvincingly excluded.

In summary, therefore, if the late onset form of AD and the inbred population study by Schellenberg et al. (1988) are set to one side, the data are consistent with the existence of a gene conferring susceptibility to AD on chromosome 21 which at present is likely to be most closely linked to the D21s16 locus. However, the evidence is based upon a small number of rather unusual autosomal dominant-like pedigrees.

Genetic Epidemiology

Having a family history of AD is virtually the only undisputed risk factor for the disorder (Heyman et al. 1984). Having said this, there is little convincing evidence purely from 'classical' studies that the aggregation of AD in families occurs because of genetic rather than environmental factors. To our knowledge there have been no adoption studies, and perhaps this is not surprising given the practical difficulties in studying a disorder of late life. Twin studies have not provided convincing evidence of genetic influences. There have been multiple reports of discordant monozygotic (MZ) pairs and, providing the unaffected co-twin survives for many years without showing signs of the disease, the existence of such pairs strongly suggests that AD is not entirely genetic in its aetiology (see recent review by Wright 1990). The only twin series of adequate size was reported by Kallmann (1956), who found a 43% concordance in MZ twins compared with 8% concordance in dizygotic (DZ) twins. There was an attempt to follow up this series by Jarvik et al. (1980) which proved to be incomplete but failed to confirm the original finding of significant differences in MZ versus DZ concordance. A more recent twin study (Nee et al. 1987) was based on a total sample size of only 22 pairs, but the MZ and DZ concordances were virtually identical at 40% and 41%, respectively.

Although, because of their small sample size, the more recent twin reports cannot exclude a genetic aetiology for AD in late life, they do make it unlikely that genes account for all of the variance in liability to develop the disorder. It is therefore surprising that some authors (Breitner et al. 1986; Martin et al. 1988) have argued that the transmission of AD, including late-onset disorder, can be explained in terms of an autosomal dominant gene which is fully penetrant. The argument put forward is that, if an age correction is performed, the cumulative risk in the first degree relatives by the age of 90 years is 50 %. While this plain man's approach to explaining the segregation of AD has a certain straightforward appeal, and would be in line with other well-documented but probably rare occurrences of large pedigrees where the disorder appears to segregate as a fully penetrant autosomal dominant character (Nee et al. 1983; Foncin et al. 1986), it overlooks the fact that the same age correction procedure results in a risk of AD as high as 20 % in control populations. We think that it is wise to exercise caution when attempting to explain the mode of transmission of a complex and difficult to diagnose trait such as AD, and to bear in mind that simple mendelian segregation ratios can be simulated by non-monogenic or indeed non-genetic familial traits (Edwards 1960). We therefore felt that it was worthwhile to explore the transmission of AD using up-to-date methods of complex segregation analysis. Unfortunately, there have been few published studies in which there has been a systematic and well-described method of ascertainment of probands, and where families are described in sufficient detail to allow segregation analysis. We have found only two such reports containing adequate sample sizes, the study of pre-senile dementia by Sjogren et al. (1952) and the study of senile dementia by Larsson et al. (1963). Although both studies are old and do not follow the modern practice of applying explicit operational diagnostic criteria, they do give adequate details of their methods, including sampling, and provide relevant population risk figures.

Segregation Analysis

The study by Sjogren et al. (1952) was of 80 probands and their families ascertained over a specific period in Stockholm and Gothenberg. Post-mortem data were available on 36 probands, of whom 18 had AD and 18 had Pick's disease. The mean age of onset in the AD group was 53 years, and it was possible on clinical grounds to define a further group of 18 probands, making a total of 36 with probable AD. These Sjogren et al. refer to as 'Group A': The study of Larsson et al. (1963) took as a starting point all cases of 'senile dementia' (corresponding clinically to what is now usually called 'senile dementia of the Alzheimer type') registered at two large Stockholm mental hospitals during a defined period. The mean age of onset was 73 years for males and 75 for females, and a subset of the sample consisting of 256 probands and their families was studied in detail in a field investigation. We performed separate complex segregation analyses on the entire material for Sjogren et al. (1952), on Sjogren's 'Group A' and on the field study sample of Larsson et al. Before going on to describe the results, we need to briefly consider the models which were applied.

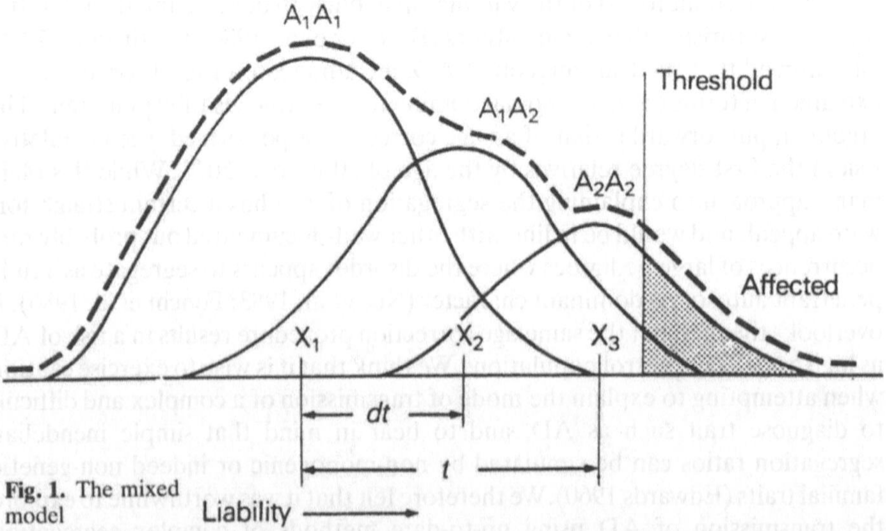

Fig. 1. The mixed model

We performed our analysis using the computer program POINTER and by applying a unified model approach (Lalouel et al. 1983), so-called because it unifies the mixed model of Morton and MacLean (1974) with the concept of transmission probabilities put forward by Elston and Stewart (1971). Under the mixed model (Fig. 1), it is assumed that a trait, x, results from the additive contributions of a major transmissible effect g, a multifactorial transmissible effect, c, and a random non-transmitted environmental component, e, such that $x = g + c + e$. The major effect results from the segregation at a single locus of two alleles A_1 and A_2 leading to three genotypes with mean liabilities of exhibiting the trait of x_1, x_2 and x_3, as shown in Figure 1. If it is assumed that liability has an overall mean of zero and a unit variance, the major gene effect is parameterised in terms of the following: q, the gene frequency of A_2; t, the displacement between x_1 and x_3, and d, a dominant parameter. The transmissible multifactorial component has a single parameter H, the proportion of total variance in liability contributed by multifactorial transmission. Under the unified model, random mating is assumed, and the parameters called tau_1, tau_2 and tau_3 denote the probabilities of transmitted allele A_1 for genotypes A_1A_1, A_1A_2 and A_2A_2 respectively. The mendelian expectations are $tau_1 = 1$, $tau_2 = 1/2$, $tau_3 = 0$. Pairwise comparison of models can be carried out using the likelihood ratio test, where minus twice the difference in log likelihoods is asymptotically a chi square with degrees of freedom (df) equal to the difference in the number of parameters.

Table 1 summarises the results of applying a mixed model to all 80 pedigrees ascertained through probands with presenile dementia in the study of Sjogren et al. (1952). The important findings are as follows. We cannot distinguish between a fully dominant gene ($d = 1$) and a more general single locus model. However, the mixed model of a major gene plus a multifactorial background offers a better explanation of the data than a single locus model alone (chi square 12.58 2 df, $p < .005$). Comparing the multifactorial and mixed models gives a negligible

Table 1. Mixed Model Analysis of Pre-Senile Dementia of All Types (Field study data of Sjogren et al (1952)

Model	d	t	q	H	-21nL+cons
mixed	1.0	0.002	0.024	0.629	-571.39
sml	0.894	1.954	0.009	[0]	-558.79
dominant	[1.0]	1.754	0.009	[0]	-558.79
multi-factorial (mf)	[0]	[0]	[0]	0.629	-571.37

likelihood ratios: mixed-sml = 12.58 (ldf); mixed-mf = 0.02 (3df)
(parameters shown in square brackets, eg [0] are fixed)

likelihood ratio chi square of 0.02 and so, on grounds of parsimony, the multifactorial model is favoured. We can next compare the multifactorial model, where presenile dementia has a heritability of about 63 %, with a 'null' model (not shown) which assumes that there are no transmissible effects. We were able to reject the null model with a likelihood ratio of 57.23 (1 df, p = 0.000). Therefore, we can arrive at two major conclusions. The first, which is not surprising, is that there is very strong evidence of transmission, but the second, which is more surprising, is that a multifactorial model offers the best explanation of the data. Since, as we have discussed earlier, it has been proposed that early onset AD is transmitted as an autosomal dominant and since it is assumed that Pick's disease is a very distinct condition also with an autosomal dominant like pattern of transmission (Sjogren et al. 1952), it might have been expected that evidence of a major gene would have arisen even from this mixed material. Since no hint of a major gene effect was obtained, we did not carry out a formal test of transmission probabilities but did go on to perform a similar mixed model analysis of Sjogren's 'Group A', the families of the probable pure AD probands. Here we obtained a flat likelihood surface and were unable, because of numerical difficulties, to obtain convergence for the mixed model. Although strictly speaking we could not compare them directly, the likelihoods for a monogenic and a multifactorial model were virtually identical. Therefore, the case for a major gene in the Sjogren 'Group A' series remains 'not proven.'

The results of the unified model analysis of the field study material of the families of probands with senile dementia (Larsson et al. 1963) are summarised in Table 2. Disappointingly, we were again unable to distinguish between the mixed model and either a single locus model (likelihood ratio chi square 0.01, NS) or a multifactorial model (likelihood ratio chi square 1.75, NS). However, there was clear evidence of a transmissible effect in that the null model (not shown in the table) provided a significantly poorer fit than the multifactorial (likelihood ratio chi square 27.42, 1 df, p = 0.000) or any of the other models. Interestingly, we were also unable to reject a single locus model when we performed the most salient test (Lalouel et al. 1983) of allowing the transmission parameter, tau_2, to deviate from its mendelian expectation. Thus a general transmission model (called the 'vertical' model in Table 2) was not significantly different from a single locus model in terms of the resultant likelihood (likelihood ratio chi square 0.53, 1 df, NS).

Table 2. Mixed Model Analysis of Senile Dementia of Alzheimer Type (Field study data of Larsson et al (1963))

Model	tau 2	d	t	q	H	21nL+cons
mixed	[.5]	1.0	1.615	0.012	0.005	–1758.55
sml	[.5]	1.0	1.619	0.012	[0]	–1758.54
multi-factorial (mf)	[.5]	[0]	[0]	[0]	0.474	–1756.80
vertical	.267	1.7	1.357	0.012	[0]	–1759.07

likelihood ratios: mixed-sml = 0.01 (ldf); mixed-mf = 1.75 (3df); vertical-sml = 0.53 (1df)
(Parameters shown in square brackets, eg [0] are fixed)

We therefore went on to estimate the probalities of genotypes in the population if we were to assume that a single gene model is 'correct.' The results of this analysis are summarised in Table 3. As we can see, the maximum likelihood single gene solution based on the field study of Larsson et al. (1963) suggests that only about 29 % of cases of AD would have at least one copy of the (A_2) mutant gene and that 71 % of cases would be true sporadics. True sporadics in the case of a monogenic disorder are those individuals who do not carry the gene but who are affected because of some other causal factors producing an indistinguishable phenotype. The use of the term 'sporadic' to denote all cases without a family history of the disorder is incorrect, since these individuals would include both true sporadics and 'chance isolates' who have inherited the gene but where cases among relatives have not been detected (Elandt-Johnson 1971).

Table 3. Probable Genotypes if the Single Major Locus Model is Correct

Probabilities (Pr)*	Genotypes		
	AA_1A_1	A_1A_2	A_2A_2
Pr (G/A)	0.71	0.284	0.002
Pr (G/N)	0.9	0.019	0.0001
Pr (G/U)	0.9	0.024	0.0001

*Probabilities of genotype given affected status where A = affected, N = not affected, and U unknown

Implications of the Genetic Evidence for an Ethical Approach to Prevention

As we have demonstrated, the genetics of AD is an extremely complex topic and there are many gaps in the available information. There is no doubt from the evidence which we have reviewed and the analyses which we have performed that both the senile and presenile forms of AD are significantly familial. This seems like an elementary point but it is important to make because there have been some recent speculations that late onset AD is so common that apparent familial clustering could be due to chance (Hardy et al. 1989). However, there is no simple explanation of an overall mode of transmission and, if anything, the genetic epidemiological evidence both in the original analysis of Sjogren et al. (1952) and in our more complicated reanalysis of their data favours a multifactorial aetiology for early-onset dementia. There are too many incriminating pieces of evidence, from observations on trisomy 21, from the location of the APP gene and from linkage studies in multiply affected families, to absolve chromosome 21 of guilt. However, is chromosome 21 or some part of its long arm the only culprit? The complicated and somewhat conflicting linkage study data certainly provoke speculation that other loci are involved. The genetic epidemiological evidence suggests that not just other loci but other factors which are not genetic in origin may be playing a part. It is therefore possible that the long arm of chromosome 21 contains not a major gene for AD in the usual sense but rather a susceptibility locus (or set of loci) which in some ways is analogous to the susceptibility loci for other common disorders such as diabetes mellitus and multiple sclerosis found in the HLA region on the short arm of chromosome 6 (Hodge and Spence 1982).

How then can our puzzling compilation of data and uncertain speculations inform an attempt to devise an ethically sound strategy for prevention? The first obvious answer is that our immediate goals should be modest. In considering genetic counselling, it is important to realise that of the three types of information listed by Murphy and Chase (1975) – empirical, modular and particular – we still only have empirical information. Even this is somewhat shaky since the risk figures in first degree relatives deriving from different family studies are based on differing diagnostic criteria and differing methods of age correction. We know that an individual's risk of AD is increased compared with the population risk if they have an affected first degree relative. We know that the risk is probably increased further if the affected relative had an early onset type of disorder and that the risk is probably increased further still if multiple relatives are affected. Beyond this it is difficult to give precise figures, and we believe that it is incorrect to attempt to provide modular information which assumes that an autosomal dominant form of AD exists with absolute certainty. For example, one of us recently was referred a young man, now in his twenties, who had a history of AD in both a parent and a grandparent and who was worried about his own declining cognitive function. Cognitive impairment was not confirmed by clinical testing but he was found to have depressive symptoms of moderate severity and concomitant difficulties in concentrating. Unfortunately, he had been told previously at another clinic that his risk of developing AD was 50 % and his worries about this clearly seemed to contribute to his depressed state.

In summary, the preferred procedure in our view is, in the face of ignorance, to admit to the uncertainties and where the choice is between providing possibly incorrect information or only limited information, to choose the latter. It is difficult to speculate on the future role of predictive testing in at-risk individuals, and we are doubtful that, even in the foreseeable future, there will be a place for prenatal testing in AD. Even if the role of linked markers on chromosome 21 can be clarified we seem to be a long way from reaching the level of precision which can be attained with chromosome 4 markers in HD. Furthermore, as we have seen even in HD, prenatal testing raises serious ethical dilemmas and is only applicable in that proportion of families who have the required informative structure.

By contrast, predictive testing in the individual at risk seems more worthy of consideration in AD. Although it is by general consent untreatable, there is evidence that at least amelioration of symptoms can be achieved with powerful inhibitors of acetyl cholinesterase, such as tacrine, and some authorities are optimistic that this approach has a promising future (Levy 1990). Since these drugs are likely to be of the greatest potential benefit in early cases, it is worth considering whether molecular genetic techniques can be utilised to produce a presymptomatic test or an aid to diagnosis in early cases. Here a susceptibility marker, even if it does not offer a high level of precision, can provide an aid in predicting risk. The obvious analogy might be with possession of a gene for hyperlipidemia in someone with a family history of coronary heart disease and the general strategy will be particularly beneficial if environmental risk factors can be identified. However, as yet such risk factors have not been definitely identified, so the analogy with coronary heart disease, where lipids can be lowered by diet or drugs, is imperfect. Clearly predictive and presymptomatic testing only has a place in subjects who specifically request such tests and who can be made aware of the implications and limitations. Furthermore, as with HD, it would seem prudent that, if and when presymptomatic tests for AD are devised, they are in the first instance offered in specialist centres which have the facilities for pretest counselling and for coping with the psychological sequelae of adverse test results.

References

Breitner JCS, Folstein MF, Murphy EA (1986) Familial aggregation in Alzheimer dementia. I. A Model for the age-dependent expression of an autosomal dominant gene. J Psychiatr Res 20: 31–43

Davies K, Read AP (1988) Molecular basis of inherited disease. IRL, Oxford

Delabar J-M, Goldgaber D, Lamour Y Nicole A, Huret J-L, Degrouchy J, Brown P, Gajdusek DC, Sinet P-M (1987) Beta-amyloid gene duplicating in Alzheimer's disease and karyotypically normal Down's syndrome. Science 235: 1390–1392

Donis-Keller H, Green P, Helms C, Cartinhour S,Weifenbach B, Stephens K, Keith TP, Bowden DW, et al. (1987) A genetic linkage map of the human genome. Cell 51: 634–641

Edwards JH (1960) The simulation of Mendelism. Acta Genet 10: 63–70

Elandt-Johnson RC (1971) Probability models and statistical methods in genetics. Wiley, New York

Elston RL, Stewart J (1971) A general model for genetic analysis of pedigree data. Hum Hered 21: 523–542

Fahy, F (1989) Different options for prenatal testing for Huntington's disease using DAN probes. J Med Genet 26: 353–603

Foncin JF, Salmon D, Bruni AC (1986) Genetics of Alzheimer's disease: a large kindred with apparent Mendelian transmission; possible implications for a linkage study. In: Briley M, Kato A, Weber M (eds) New concepts in Alzheimer's disease, Macmillan, London, pp 242–256

Gardiner K, Watkins PC, Munke M, Drabkin HA, Jones C, Patterson D (1988) A partial physical map of human chromosome 21. Somatic Cell Mol Genet 14: 633–638
Goate AM, Hardy JA, Owen MJ (1989) The genetic aetiology of Alzheimer's disease. Int Rev Psychiatry 1: 243–248
Goldgaber D, Lerman ML, McBride OW, Sffiotti U, Gajdusek DC (1987) Characterisation and chromosomal localisation of cDNA encoding brain amyloid of Alzheimer's disease. Science 235: 877–880
Gusella JF, Wexler NS, Conneally PM, Naylor SL, Anderson A, Tanzi RE, Watkins PC, Ottina K, Wallace MR, Sakaguchi AY, Young AB, Shoulson I, Bonilla E, Martin JB (1983) A polymorphic DNA marker genetically linked to Huntington's disease. Nature 306: 234–238
Hardy JA, Goate AM, Owen MJ, Mullian MJ, Rossor MN, Pearson RCA (1989) Modelling the occurrence of pathology of Alzheimer's disease. Neurobiol Aging 10: 429–431
Harper PS (1988) Practical genetic counselling, 3rd edn Wright, London
Harper PS, Morris MJ (1990) Genetic testing for Huntington's disease. (Editorial). Br Med J 300: 1089–1090
Heston LL, Mastri AR (1977) The genetics of Alzheimer's disease. Associations with haemotologic malignancy and Down's syndrome. Arch Gen Psychiatry 34: 396–406
Heyman A, Wilkinson WE, Stafford JA, Heims MJ, Sigmon AH, Weinberge T (1984) Alzheimer's disease. A study of epidemiological aspects. Ann Neurol 15: 335–341
Hodge SE, Spence MA (1982) Some epistatic two locus models of disease II. The confounding of linkage and association. Am J Hum Genet 33: 396–406
Jarvik LF, Ruth V, Matsuyama SS (1980) Organic brain syndrome and aging: a six year follow-up of surviving twins. Arch Gen Psychiatry 37: 280–286
Kallmann FJ (1956) Genetic aspects of mental disorders in later life. In: Kaplan OJ (ed) disorders in later life, 2nd edn. Oxford University Press, London, pp 24–46
Kraepelin E (1922) Manic depressive insanity and paranoia. Livingstone, Edinburgh
Lalouel DM, Rao DC, Morton ME, Elston RC (1983) A unified model for complex segregation analysis. Am J Hum Genet 35: 816–826
Larsson T, Sjogren T, Jacobson G (1963) Senile dementia: a clinical, sociomedical and genetic study. Acta Psychiatr Scand [Suppl 167] 39: 1–259
Levy R (1990) Controversies in therapeutics: are drugs targetted at Alzheimer's disease useful? Useful for what? Br Med J 300: 1131–1132
Martin RL, Gerteis G, Gabrielli WF, Martin WF (1988) A family-genetic study of dementia of Alzheimer type. Arch Gen Psychiatry 45: 894–900
McGuffin P, Sargeant M (1989) Ethics and psychaitric genetics. Curr Sci 2: 681–684
Meyer JE (1988) The fate of the mentally ill in Germany during the Third Reich. Psychol Med 18: 575–581
Morris MJ, Tyler A, Harper PS (1988) Adoption and genetic prediction for Huntington's disease. Lancet 2: 1069
Morris MJ, Tyler A, Lazarou L, Meredith L, Harper PS (1989) Problems in genetic prediction of Huntington's disease. Lancet 1: 601–603
Morton NE, Maclean CJ (1974) Analysis of familial resemblance. III. Complex segregation analysis of quantitative traits. Am J Hum Genet 26: 489–503
Murphy EA, Chase GA (1975) Principles of genetic counselling. Year Book Medical Publishers, Chicago
Nee LE, Polinski RJ, Eldridge R, Weingartner H, Smallberg S, Ebert M (1983) A family with histologically confirmed Alzheimer's disease. Arch Neurol 40: 203–208
Nee LE, Eldridge R, Sunderland T et al. (1987) Dementia of the Alzheimer type: clinical and family study of 22 twin pairs. Neurology 37: 359–363
Owen MJ, James LA, Hardy JA, Williamson R, Goate AM (1990) Physical mapping around the Alzheimer disease locus on the proximal long arm of chromosome 21. Am J Hum Genet 46: 316–322
Pericak-Vance MA, Yamaoka LH, Haynes CS, Speer MC, Haines JL, Gaskell PC, Hung W-Y, Clark CM, Heyman AL, Tratter JA, Eisenmenger JP, Gilbert JR, Lee JE, Alberts MJ, Dawson DV, Bartleit RJ, Earl NL, Siddique T, Vance JM, Conneally PM, Roses AD (1988) Genetic linkage studies in Alzheimer disease families. Exp Neurol 102: 271–279

Podlisny MB, Lee G, Selkoe DJ (1987) Gene dosage of the amyloid beta precursor protein in Alzheimer's disease. Science 238: 669–671

Quarrell OWJ, Meredith AL, Tyler A, Yongman S, Upadhyaya M, Harper PS (1987) Exclusion testing for Huntington's disease in pregnancy with a closely linked DNA marker. Lancet 1: 1281–1283

Roll-Hansen N (1988) The progress of eugenics: growth of knowledge and change in ideology. Hist Sci 26: 295–331

Schellenberg GD, Bird TD, Wisjman EM, Moore DK, Boehnke M, Bryant EM, Lampe TH, Nochlin D, Sumi SM, Deeb SS, Beyreuther K, Martin GM (1988) Absence of linkage of chromosome 21q21 markers to familial Alzheimer's disease. Science 241: 1507–1510

Sjogren T, Sjogren H, Lindgren AGH (1952) Morbus Alzheimer and Morbus Pick. A genetic, clinical and patho-anatomical study. Acta Psychiatr Scand [Supp] 82

St. George-Hyslop PH, Tanzi RE, Polinsky RJ, Haines JL, Nee L, Watkins PC, Myers RH, Feldman RG, Pollen D, Drachman D, Growdon J, Bruni A, Foncin J-F, Salmon D, Fromelt P, Amaducci L, Sorbi S, Oiacentini S, Stewart GD, Hobbs WJ, Conneally M, Gusella JF (1987) The genetic defect causing familial Alzheimer's disease maps on chromosome 21. Science 235: 885–890

Tanzi RE, Gusella JF, Watkins PC, Bruns GAP, St. George-Hyslop, PH, Van Keurin ML, Patterson D, Pagan S, Kurnit DM, Neve RL (1987) Amyloid B protein gene: cDNA, mRNA distribution, and genetic linkage near the Alzheimer locus. Science 235 (4791): 880–884

Tanzi R, Haines J, Watkins P, Stewart G, Wallace R, Hallewell R, Wong C, et al. (1988) Genetic linkage map of human chromosome 21. Genomics 3: 129–136

Volkers WS, Went LN, Vetger van der Lis M, Harper PS (1980) Genetic linkage studied in Huntington's Chorea. Annals of Human Genetics 44: 75–79

Warren AC, Robakis NK, Ramakrishna N, Koo EH, Ross CA, Robb AS, Folstein MF, Price DL, Antonarakis SE (1987) Beta-amyloid gene is not present in three copies in autopsy validated Alzheimer's disease, Genomics 1: 307–312

Weatherall D (1985) The new genetic and clinical practice, 2nd edn. Oxford University Press, Oxford, p. 206

Whalley WL, Carothers AD, Collyer S, De Mey R, Frackiewicz A (1982) A study of familial factors in Alzheimer's disease. Br J Psychiatry 140: 249–256

Wright AF (1990) The genetics of common forms of dementia. In: McGuffin P, Murray RM (eds) The new genetics. Heinemann, London

Some Ethical Aspects of Genetic Aproaches to Human Health Care: A Developmental Geneticist's Point of View

I. Ivanov

Indroduction

There are distinct dissimilarities between phenomena taking place at different levels of life: what is true for cellular macromolecules may not be quite true for cellular processes as a whole, and may be still less true for the ontogeny of an organism. For example, the severity or mildness of such genetic diseases as thalassaemias is not directly related to the size of respective genetic lesions. In the course of phenotypic realization of genetic information in development, many enhancers and inhibitors can and do interfere with the process, so the connection between the genetic cause and the phenotypic effect may become rather equivocal.

Metazoan development, including man, comprises a complex network of ultracellular, cellular, and infracellular phenomena of growth, form-building, and function establishment which are perfectly coordinated both spatially and temporally. They ensure the transformation of a single omnipotent initial cell, the fertilized egg-cell, into a multicellular organism exerting species-specific, family-specific and individual features of its morphological, physiological and, in the case of man, psychological habitus.

It is generally believed that the stability and reproducibility of organism development are achieved through genetic mechanism. This suggestion seems quite plausibel; a great number of genetic mutations are known to be capable of altering respective steps in organism development. Some genetic mechanisms of the control of respective developmental events have already been revealed.

However, since any structural gene can fulfil only two simple functions, namely to replicate its own DNA sequence and to transcribe its deoxyribose chains into polyriboside RNA sequences, it is to be inferred that coordinated genetic regulation of organism development is carried out at the systemic level, involving a network of control circuits consisting of both negative and positive, inductive and repressive feedback mechanism.

It is generally appreciated that the greater the numbers of structural components and functional steps in the process of information transfer are, the "louder" is the noise accompanying it. Therefore, more or less deterministic cause and effect relations can only be traced to primary gene products. And the more remote the traits under consideration are from the primary gene expression, the more equivocal are the genes to effect relations.

Srám, Bulyzhenkov, Prilipko, Christen (Eds.)
Ethical Issues of Molecular Genetics in Psychiatry

Some genes (their alleles or mutants, to be exact) exert particularly strong or even crucial effects on respective hereditary traits, resulting in true Mendelian segregation in progeny. By 1989, some 4,200 Mendelian characteristics were catalogued by Professor V.A. McKusick. More often, however, especially when behavioural characteristics are concerned, significant deviations arise from the expected Mendelian ratios due to incomplete penetrance and/or variable expressivity of respective genes and genotypes. Actual values of the penetrance and expressivity depend on cooperative action of genetic milieu with the complex of intraorganismal and external factors of development. As a result, the "genes-effects" relations are usually probabilistic rather than deterministic in case of sophisticated physiological and psychological end-points.

It seems that there are sound reasons to account for the systemic action of multicomponent gene complexes and variability in gene expression when considering the ethical issues of application of genetic approaches to health care problems. And this is particularly important when mental disorders are concerned, since the majority of the latter comprise multifactorial conditions determined by complicated interactions of genetic and environmental factors.

Ethics and Genetics

Irrespective of the kind of humanistic problems in question, the general ethical principles are essentially the same. These include (Macklin 1989):

(a) "the principle of individual liberty: individuals have a right to freedom of decision and action to the extent that their actions do not interfere with the rights of others.

(b) the utilitarian principle: morally right actions or policies are those that result in the greatest good for the greatest number of people.

(c) a principle of justice: all persons within a given society deserve equal access to goods and services that fulfil basic human needs".

Within the realm of the present topic one also has to bear in mind some more specific consideration concerning the genetic load of mankind and some current genetic approaches to health care problems. Humanity is overloaded with genetic and related disorders (mental disorders among them), and we have to do something for the good of the patient, the family and the population as a whole by means of prevention or treatment of the diseases.

Molecular genetic approaches, DNA technology in particular, open the way to provide for early presymptomatic and prenatal identification of certain gene carriers or affected individuals, as well as those with later onset of the disease.

These molecular genetic achievements enable us to foresee that current means of symptomatic and pathogenetic treatment of hereditary diseases can be essentially strengthened, if not replaced, by substitutional gene therapy in which an artificial recombinant DNA construction will be integrated into respective genomes to restore their normal function. Currently such etiological treatment of genetic disorders is not available practically. Therefore, common medical genetic practice is based on early prenatal examination of fetus health with consequent artificial termination of pregnancies at risk of delivering an ill or malformed child.

Here again some more ethical problems of choice of decision unavoidably arise. One of particular importance was formulated by Theodosius Dobzhansky in the following way: "If we enable the weak and the deformed to live and promulgate their kind, we face the prospect of a genetic twilight. But if we let them die or suffer when we can help them, we face the certainty of moral twilight."

The policy of elective abortion of genetically affected fetuses also raises a number of other related decision-making dilemmas. For example, whose judgment is to be considered reliable to determine the adequate proportions of different genotypes in a population? How many Einsteins, Leonardos and Beethovens does a human population actually need? And is it not more just to declare and to follow strictly the principle of equal rights to life and equality of all fetuses, except for inviable and severely malformed or incurably ill ones?

If and when we resolve these controversies, we will still have to deal with further ethical problems inherent to molecular genetics itself. Such problems arise from the basic features of molecular genetic applications to disease prevention and treatment, namely:

1. in contrast to approaches based on examination of clinical phenotypes, the molecular genetic approach to disease prevention and treatment involves studies of neither biochemical and physiological manifestations of gene action nor or the definitive structural, physiological and behavioural traits, but rather of the fine structure of respective DNA sequences, i.e., the genetic substance itself is examined or artificially altered.
2. with an increase in the sensitivity and precision of molecular genetic methods, progressively smaller and smaller samples of DNA material become sufficient to perform sophisticated qualitative and quantitative investigations into the genetic fine structure. On the other hand, in any molecular genetic manipulation only a very small fraction of genomic DNA is subjected to either analytical or recombinatorial procedures.
3. in the case of mental disorders there is always a rather long period between the early presymptomatic diagnosis and the actual late onset of the disease.
4. effective methods are already available, at least for laboratory organisms, to integrate the exogenous and/or recombinant DNA into both the somatic and germ cell genomes and to enable its replication and transmission to succeeding cell and/or organism generations.

The above features of molecular genetic technology set certain limitations on the application of molecular genetic approaches to disease prevention and treatment, and most of these limitations are ethical rather than technological in their essence.

The molecular genetic approach is especially useful in the prevention of strictly Mendelian conditions, since it is entirely unable to account for variations in manifestation and thus avoid false-positive and false-negative decisions in cases of poorly segregating conditions. Poor and variable segregation ratios are typical for psychiatric discorders; therefore, an ethical problem arises of communication of wrong results to the patient and the family, as well as of elimination of actually unaffected fetuses.

A similar problem is characteristic of gene therapy, which would insert supposed to be good DNA sequences into an unknown genetic environment with subsequent propagation of the DNA in randomly varying genetic background. How can one be sure that the inserted DNA sequence will unmistakably be to the good of a recipient, as well as to all of the descendants who inherit the sequence from him or her?

The very high resolution power of the methods of molecular analysis permits backtracking the time of the biopsy sampling to the earliest gestational ages (if not to prognesis itself). This makes molecular genetic analysis of the conseptuses DNA in order to make a decision on elective abortion more acceptable from both the legal and ethical points of view. However, some orthodox Catholics (Professor Jerome Lejeunne among them) claim that only the prefertilization gametes should be considered as parts of the bodies of their producers, who are within their full rights to dispose of them at their free will. At the moment of fertilization, however, the zygote immediately becomes an animated individual possessing all the legal rights attributable to a human being. This concept is disputable, indeed, but it does exist and has a sound basis.

Finally, the judgment on an actual value of a human genome based on the analysis of its minute fractions is an oversimplification. How can one be sure that, when eliminating a supposed to be "bad" gene, a precious genome to which the "bad" gene belongs is not simultaneously thrown away? How does one predict unequivocally the manifestation of the "bad" gene when it is present in the genome in combination with some other modifier genes? Is it not safer and ethically more sound to restrict the therapeutical genetic manipulations to somatic cells only?

Conclusions

Molecular genetic achievements may be useful in the identification and genomic localization of DNA sequences relevant to the origin of diseases, including psychiatric disorders.

DNA technology may serve as a useful tool in planning reproduction policy for families at risk.

References

Bankowski Z, Barzellato J, Capron AM (eds) Ethics and human values in family planning. CIOMS, Geneva, pp 68–85

Macklin R (1989) Perspectives of Different Cultural and Religious Testing. In: Bankowski Z, Barzelatto J, Capron AM (eds) Ethics and human values in family planning. CIOMS, Geneva, pp 68–85

Lessons from an International Survey of Medical Geneticists

D.C. Wertz

Overview

This paper focuses on selected results of a 1985–86 survey of ethical decision-making among 682 geneticists in 19 nations. The entire survey has appeared elsewhere (Wertz and Fletcher 1989, Wertz et al. 1990); in what follows I have selected those issues most germane to psychiatry. A few non-psychiatric examples, such as population screening for cystic fibrosis, have been included, because these represent generic issues (e.g., mandatory versus voluntary screening) that can also be applied to psychiatry.

In a previous paper, presented last year at the WHO/IPSEN Foundation conference on "Genetic Approaches in the Prevention of Mental Disorders" (Wertz 1990), I outlined specific ethical problems and suggested resolutions which I will not repeat in detail here. The present paper, which focuses on survey results, should be read in conjunction with the previous paper (Wertz 1990), which focuses on policy. I begin and end with some references to history as a necessary warning.

An Historical Caveat

Explanations for mental illnesses, behavioral disorders, and crime have followed historical cycles that alternate between biological and social/environmental etiologies. From the beginning of this century until World War II, biological explanations were in ascendance. In the United States, during these years, family studies of rural whites with low I.Q's or criminal temperaments, such as "Joe Sixty" (so named for his I.Q), were used to institute measures of social control (sterilization laws in 30 states, forced institutionalization) over unruly segments of the population and to bolster the credibility of the new professions (psychiatric social workers, social psychologists, criminologists) responsible for placing such people in public institutions (Fink 1938; Chorover 1979; Kevles 1985; Rafter 1989). Prevention meant negative eugenics; treatment was essentially nonexistent, but meant separating deviants from society through long-term incarceration.

Biological theories of the origins of crime had important impacts on general medicine in the United States. For example, the routine episiotomies and outlet forceps deliveries used in American births for almost 70 years resulted from

Srám, Bulyzhenkov, Prilipko, Christen (Eds.)
Ethical Issues of Molecular Genetics in Psychiatry

doctors' beliefs that surgically unassisted births so damaged the baby's brain as to predispose it to criminal behavior (DeLee 1920; Wertz and Wertz 1989).

Most of these biological theories were scientifically sloppy. At their worst, they credited environmental factors to heredity. For example, Charles B. Davenport, director of the Cold Spring Harbor's Eugenics Record Office, claimed to have found a heritable trait called "thalassophilia" or "love of the sea" among the families of naval officers. He noted that because it appeared almost exclusively in males, it must be sex-linked, and because it sometimes skipped generations, it must be a Mendelian recessive (Davenport 1919).

Biologically based theories of behavior did not die with the discreditation of Nazi eugenic experiments. In fact, the most expensive medical experiment ever undertaken in the United States before 1960 involved a positive eugenic approach to temperament. This was the Rockefeller Foundation's attempt, in the 1940s and 1950s, to breed the ideal dog (Paul 1991). The dog experiment was designed to demonstrate to the general public the heritability of intelligence and emotion and to convince people that selective breeding could make a difference in their own lives. Sales of these "superdogs" were to pay for the experiment. In the end, the experiment failed, leaving its director to wonder whether any aspects of temperament were heritable.

Biological theories of behavior slowly lost scientific credibility in the 1950s and were replaced by social, psychological, and environmental approaches. Behavior was seen as the result of poverty, race or class discrimination, broken families, poor schools, or poor neighborhoods. Prevention meant large-scale social programs designed to combat poverty; treatment meant individual or family psychotherapy or short-term stays in rehabilitative institutions. Hundreds of thousands of mental patients were released from long-term institutions and placed in what was supposed to become a network of "community care". All these programs were expensive; many were never carried through. The large-scale social/class changes that would have been necessary to support environmental treatment were not possible politically. Traditional psychotherapy appeared to offer, for the few who could afford it, at best a one-third chance of cure. By the 1970s, the public began to perceive social approaches to mental illness and crime as failures.

New biological approaches, stemming from greater knowledge about genetics, appeared to offer the best hope to the mentally ill, who were for the first time organizing themselves to fight discrimination and to demand more services. By claiming that mental illness was genetic in origin, patients and their families felt that they could remove the stigma attached to their condition, just as Alcoholics Anonymous had done by making alcoholism a disease. A genetic origin for mental illness also removed the blame from patients' families, who were tired of being labeled "schizophrenogenic mothers" or "emotionally distant fathers". The National Association of the Mentally Ill, a grass-roots organization of patients and families, wholeheartedly supported genetic approaches over social/environmental approaches. Another attractive aspect of the genetic approach was that it provieded hope for a definitive, permanent treatment through somatic-cell or germ-line therapy, though such treatments are far in the future.

This is the "era of genetics" in psychiatry. How long it will last and what preventive measures or treatments will succeed remain to be seen. We should not forget that biologically grounded theories have been in ascendance before. Such theories tend to enjoy favor during conservative periods because they do not require painful social changes and they also provide a rationale for social control through some form of negative eugenics, such as birth control, sterilization, or (now) prenatal diagnosis and selective abortion. Although today's science of genetics is on firmer ground than past biological theories in psychiatry, in the end it will probably fail to explain the majority of cases of deviant behavior (Holtzman 1989; Billings 1990). In supporting genetics for the good that will result to some patients, we should not lose sight of the larger social programs that will be necessary to prevent and treat most mental disorders.

Problems Arising from Incomplete Science

There are two basic underlying problems in genetic approaches to psychiatry. The first is diagnosis in the absence of treatment. The second is the extremely varied nature of the conditions commonly labeled "mental disorders".

Although the long-term goal of genetic research is treatment, for the foreseeable future most genetics will be diagnostic. Even after the human genome is mapped and sequenced, few treatments will result from this enormous expenditure unless governments permit research on the human embryo. Patients may face a long period of ever-more-precise diagnosis but no treatments. "Prevention" will mean negative eugenics: refraining from having children, using someone else's sperm or egg, or using prenatal diagnosis and selective abortion. For some disorders, prevention could also mean taking early social/environmental measures to ameliorate the effects of the condition (e.g., special early schooling for dyslexics or XYY), but the effectiveness of such preventive measures has not been demonstrated. For some single-gene disorders, such as Huntington's disease, no environmental prevention is possible. There is a strong possibility that for many disorders societies will find prevention – through negative eugenics – less expensive than research on genetic treatment, and that genetic studies will not progress beyond prenatal diagnosis.

Psychiatry commonly includes a wide variety of conditions under "mental disorders". These range from mental retardation to affective disorders (schizophrenia, bipolar disorder), learning disabilities (dyslexia), neuroses, and behavioral disorders (drug addiction, alcoholism, crime). Some conditions are totally genetic in origin (e.g., Down syndrome, Huntington's disease, possibly ten percent of Alzheimer's disease). Most disorders are multifactorial, involving a combination of several genes plus environment. Others may be totally environmental. In multifactorial disorders, it is usually impossible to determine what proportion of etiology is genetic and what environmental. Individuals at risk for multifactorial disorders may be given risks based on family history (and, in some cases tests), but most people find these risks difficult to interpret, partly because so much depends on the environment. In the case of children, it is questionable whether early diagnosis helps or hinders. For example, a child with XYY may receive special schooling and parental treatment that prevents him from becoming violent, but in the process he is stigmatized and may develop other psychiatric problems. The peculiarly stigmatizing nature of mental disorders, and the unknown proportion of genetic/environmental etiology in most, complicate the already delicate ethical issues in human genetics.

Major Ethical Problems in Genetics: The 19-Nation Survey

Most of the major ethical problems of genetics in psychiatry are the same as those for the wider field of human genetics in general; disclosure, directive versus nondirective counselling, mandatory versus voluntary screening, and indications for prenatal diagnosis. To provide a basis for international discussion, in 1985–86 Dr. John C. Fletcher and I undertook a survey of geneticists' responses to ethical dilemmas in 19 nations, using anonymous questionnaires distributed through a contact geneticist in each nation. Copies of the questionnaire appear in Wertz and Fletcher (1989; Wertz et al. 1990). On the basis of Dr. Fletcher's fieldwork at genetics centers in 12 nations, we constructed 14 case vignettes and four screening or testing situations describing the most difficult

Table 1A. Nations participating in 1985–86 survey of ethics and genetics

Nation	Number asked to participate	Number of persons responding	Response rate (%)
Australia	14	12	86
Brazil	51	32	63
Canada	73	47	64
Denmark	28	15	54
Federal Republic of Germany (FRG)	55	47	85
France	35	17	49
German Democratic Republic (GDR)	25	21	84
Greece	11	7	64
Hungary[a]	18	15	83
India	72	27	38
Israel	17	15	88
Italy	26	11	42
Japan	74	51	69
Norway	10	6	60
Sweden	26	21	81
Switzerland	10	5	50
Turkey	13	5	38
United Kingdom	50	33	66
United States	490	295	60
Total	1.098	682	62

[a] Unit of analysis is genetics centers

cases, ethically speaking, frequently occurring in genetics. For each case, we asked respondents what they would do (from a checklist of possible actions) and why, in their own words, they would do it. We also included 12 questions on goals and approaches in counselling, based on a 1977–79 survey in the U.S. (Sorenson et al. 1981). In each nation, we asked our contacts to include all medical geneticists at the M.D. or Ph.D. level, including laboratory scientists, administrators, physicians, and counsellors. We excluded Masters'-level genetic counsellors because this newly emerging profession is found mostly in the United States and Canada. Of 1098 asked to participate, 682 (62 %) responded by the close of the study in early 1987 (Table 1). In the United States, a comparison between respondents and nonrespondents by type of degree, speciality in genetics, gender, and geographical area revealed no statistically significant differences between the two groups. This suggests that, at least in the U.S., respondents were representative of genetic professionals.

Three of our 14 case vignettes, two screening or testing situations, and all 12 counselling questions have direct relevance to genetics in psychiatry. These are reported below.

Table 1B. Professional Characteristics of geneticists responding to survey

Country	Total % M.D.'s	% Pediatricians	% Ph.D.'s	Median years in genetics	Median hours/wk in genetics	Median patients/wk counseled
Australia	100	75	0	11	40	8
Brazil	56	38	38	13	40	4
Canada	72	38	26	15	45	3
Denmark	100	20	0	17	40	3
Federal Republic of Germany	96	23	4	9	39	7
France	100	88	0	17	34	12
German Democratic Republic	95	48	0	10	40	8
Greece	71	57	14	15	38	8
Hungary	100	53	0	18	17	5
India	79	48	19	7	39	4
Israel	67	40	20	11	42	8
Italy	100	36	0	10	31	7
Japan	100	55	0	10	20	1
Norway	100	0	0	9	49	4
Sweden	90	43	10	8	38	4
Switzerland	80	20	20	13	41	4
Turkey	80	20	20	9	41	8
United Kingdom	91	27	9	13	39	9
United States	74	51	22	14	45	4
Total	81	45	16	14	45	4
Total, excluding U.S.	86	39	12	12	40	5

Disclosure of Psychiatric Diagnoses

Sometimes a patient is ashamed of a diagnosis or fears that the news will leak out to an employer or insurer, with likely loss of employment or insurance resulting. While still mentally competent, some patients with Huntington's disease (HD) refuse to disclose, or permit the doctor to disclose, the diagnosis to spouse or relatives at genetic risk of developing HD. We posed the following question:

"A client recently diagnosed as having Huntington's disease refuses to permit disclosure of the diagnosis and relevant genetic information to siblings who may be at risk of Huntington's disease. In your professional capacity as a medical geneticist, what would you do?" The results indicated no consensus among geneticists (Table 2): 32 % would maintain the patient's confidentiality, 34 % would tell the relatives if they asked, 24 % would seek out and tell the relatives even if they did not ask, and 10 % would refer the matter back to the family doctor for a decision. Those who would refer argued that the family doctor best knows how and when to tell the family, or (in Japan) that the family doctor has

Table 2 Patient confidentiality versus duties to third parties. Percent of respondents who would disclose diagnosis to relatives at risk, against patient's wishes.[a]

Country	Huntington's disease	
Australia	75*	(0)+
Brazil	91*	(38)
Canada	57	(32)
Denmark	87*	(33)
Federal Republic of Germany	74	(9)+
France	47	(18)+
German Democratic Republic	57	(43)
Greece	14+	(0)+
Hungary	47	(40)
India	48	(33)
Israel	73	(27)
Italy	82*	(36)
Japan	53	(16)+
Norway	83*	(50)
Sweden	48	(14)+
Turkey	80*	(60)
Switzerland	40	(20)+
United Kingdom	58	(32)
United States	53	(24)+
Total	58	(24)+
(95 % confidence limits)	(53, 63)	(20, 28)
Total, excluding U.S.	63	(25)+
(95 % confidence limits)	(57, 69)	(20, 30)

[a] In parentheses, percent who would diclose even if relatives did not ask.
* Strong consensus for disclosure;
+ Strong consensus against disclosure

the authority to make the patient disclose. Respondents found this the second most difficult of the entire 14 cases in the questionnaire (after prenatal diagnosis for sex selection). A question on disclosure of results of presymptomatic testing for HD showed similar absence of consensus about telling relatives or spouse (Table 3). In Greece, India, and Turkey, where most would tell the spouse, HD is rare and many geneticists have never seen a case. This may have affected their responses. The overwhelming majority (93–96 %) would not reveal test results to employer, insurer, or school without the patient's consent, and substantial

Table 3. Access to results of presymptomatic tests for Huntington's disease

Country	Percent responding "No access to results without patient's consent"[a]				
	Patient[++]	Spouse	Relatives at risk for HD	Employer	Life and medical insurers
Australia	50	75* (8)	67	100* (67)	100* (58)
Brazil	56	44 (6)	16[+]	94* (31)	91* (28)
Canada	67	56 (7)	65 (10)	100* (54)	100* (46)
Denmark	73	53 (7)	27	100* (60)	100* (67)
Federal Republic of Germany	84* (4)	82* (7)	69 (4)	100* (91)*	95* (84)*
France	71 (6)	77* (15)	46 (8)	100* (83)*	100* (67)
German Democratic Republic	75*	79* (16)	65 (20)	100* (89)*	90* (85)*
Greece	33	17[+]	50*	100* (17)	83* (17)
Hungary	53 (13)	67 (20)	27 (7)	93* (60)	87* (60)
India	30 (4)	24[+] (4)	27 (4)	70 (35)	60 (40)
Israel	57	50	46 (8)	100* (50)	100* (21)
Italy	67	50 (10)	50 (10)	100* (60)	100* (50)
Japan	57 (8)	55 (17)	37 (14)	85* (67)	88* (70)
Norway	50	67	50	100* (83)*	100* (83)*
Sweden	71	79*	57	100* (62)	100* (48)
Switzerland	80*	60	80*	100* (80)*	100* (60)
Turkey	0[+]	0[+]	0[+]	50 (25)	75*
United Kingdom	94*	77*	71	97* (16)	100* (19)
United States	67	65 (1)	57 (2)	98* (34)	94* (36)
Total (95 % confidence limits)	66 (2) (61, 71)	62 (5) (57, 67)	52 (4) (47, 57)	96* (46) (94, 98)	93* (45) (90, 96)
Total, excluding U.S. (95 % confidence limits)	65 (3) (59, 71)	61 (7) (55, 67)	48 (5) (42, 54)	95* (56) (92, 98)	93* (52) (90, 96)

[a] In parentheses, percent responding "No access at all".
[++] Those who would tell patients only if they want to know, as opposed to informing patients against their will.
* Strong ($\geq$ 75 %) consensus in favor of access only with consent.
[+] Strong ($\geq$ 75 %) consensus for access without consent.

minorities (45–46 %) would not allow these institutional third parties access to test results even with the patient's consent.

Although HD is a single-gene disorder, it can be used as a paradigm for disclosure in other disorders with a genetic component. Dr. Fletcher and I have argued that, because genetic disorders are familial, the knowledge of diagnosis should not be the exclusive property of one individual. Family members have a moral obligation to reveal potentially useful genetic information to each other and to participate in research and testing that will help other family members. If children are contemplated, patients also have a moral obligation to tell their spouses so that both may make a decision that will be in the best interests of the potential child. (Even if patient and spouse are beyond childbearing age, the patient has a moral obligation to tell the spouse, who will probably be responsible for her/his care). Geneticists have a professional duty to tell patients of the patient's moral obligation to reveal certain genetic information to spouse and family. This is not a situation for "nondirective" counselling. If a couple is fertile, some geneticists have proposed including a provision in the informed consent agreement that the patient either tell the spouse or allow the doctor to do so, and would withhold testing if the patient refused (Holtzman 1989). Withholding testing appears to be a coercive measure, and could set a dangerous precedent. We do, however, believe that doctors should be legally permitted to tell fertile spouses and family members at genetic risk if the patient refuses to do so. The underlying ethical principle is avoidance of harm to innocent third parties. In the United States, the President's Commission (1983) has set forth the following four conditions under which a patient's confidentiality may be breached in order to warn third parties:

> "A professional's ethical duty of confidentiality to an immediate patient or client can be overridden only if several conditions are satisfied: [1] reasonable efforts to elicit voluntary consent to disclosure have failed; [2] there is a high probability both that harm will occur if the information is withheld and that the disclosed information will actually be used to avert harm; [3] the harm that identifiable individuals would suffer would be serious; and [4] appropriate precautions are taken to ensure that only the genetic information needed for diagnosis and/or treatment of the disease in question is disclosed" (p.44).

Although we agree that doctors should be *permitted* to disclose in these cases, we do not advocate that they be legally *required* to do so. A legal requirement to disclose would place an unfair burden on the doctor, who might be unable to locate relatives at risk. Delegating the responsibility to the government health department, by requiring the physician to report genetically transmissible disorders to public health officials (as is now required for sexually transmissible diseases such as syphilis and AIDS), who then seek out and inform third parties at risk, would be inadvisable, because it would place too much personal information in government hands, with the possibility of discrimination and eugenic abuses. Although genetic disorders are transmissible to the next generation, they should not be treated like communicable diseases such as typhoid or AIDS.

Protection of confidential genetic information from employers and insurers will be difficult, perhaps impossible in many nations, unless a convincing case can be made that genetic information differs from other medical information. A better means of protecting individuals from loss of job and insurance is to place genetic or partially genetic disorders under disability-rights legislation that prohibits discrimination in employment as long as the worker is able to perform the job and that requires insurers to provide health coverage to all for a reasonable fee. The latter could be extremely costly for private insurance companies in countries without national health insurance, and could lead these countries to replace private insurance systems with government-supported health insurance. Life insurance – not a basic human right – is a separate issue that should not be confused with health insurance. Life insurance is basically a commercial enterprise, and life insurers should be allowed access to medical/genetic information affecting longevity, but they should be required to provide some coverage – excluding the risked condition, if desired.

In jobs involving the safety of others, the public interest will be best served by regular testing and monitoring of functional abilities and mental condition for all employees, taking into account their medical histories and any history of substance abuse. It would be unwise and unfair to single out for special surveillance those with genetic risks, because most accidents resulting from human failures are unrelated to genetics. Alcohol, drug abuse, and sleep deprivation play a far greater role in occupational accidents than disorders of genetic origin.

Directive Versus Nondirective Counselling

Disclosure is usually followed by genetic counselling. The gospel of nondirectiveness in counselling has spread around the world (Wertz and Fletcher 1988) with a few exceptions, such as Hungary (Czeizel 1989) and India. Tables 4A and 4B report responses to our 12 questions on goals and methods of counselling and compare these responses with a 1977–79 survey in the United States (Sorenson et al. 1981). Most geneticists believed that the primary goals of counselling were to provide information that would help patients make informed decisions, help patients cope with genetic problems, remove patient's guilt or anxiety, and help patients achieve their parenting goals (Table 4A). Although these four educationally oriented goals enjoyed the strongest consensus, three goals that were more eugenically oriented also attracted some support. In all, 97 % thought that "prevention of disease or abnormality" was an important goal of counselling, with 31 % considering it "absolutely essential". A total of 74 % thought "improvement of the general health and vigor of the population" (positive eugenics) was important, and 14 % considered it "absolutely essential". Although there was less support for the most openly eugenic goal, "reduction of the number of carriers", 54 % considered it important and 5 % essential.

With regard to approaches (Table 4B), most (94 %) said that they would tell patients that they would support any decisions the patient made, and 92 %

Table 4A. Seven Goals of Counselling. Percent believing important[a]

Country	Goal: Helping individuals/couples understand their options and the present state of medical knowledge so they can make informed decisions		Helping individuals/couples adjust to and cope with their genetic problems		The removal or lessening of patient guilt or anxiety		Helping individuals/couples achieve their parenting goals		The prevention of disease or abnormality		Improvement of the general health and vigor of the population	
Australia	100*	(100)*	100*	(100)*	100*	(50)	100*	(33)	92*	(17)	67	
Brazil	100*	(84)*	100*	(81)*	100*	(78)	97*	(41)	100*	(66)	72	(25)
Canada	100*	(98)*	100*	(87)*	100*	(57)	98*	(41)	98*	(11)	68	(11)
Denmark	100*	(100)*	100*	(100)*	100*	(60)	92*	(46)	100*	(53)	86*	(14)
Federal Republic of Germany	98*	(89)*	100*	(89)*	100*	(75)	98*	(43)	91*	(23)	28	(4)
France	100*	(94)*	100*	(88)*	100*	(69)	94*	(63)	100*	(69)	81*	
German Democratic Republic	100*	(57)	100*	(86)*	100*	(38)	100*	(24)	100*	(76)*	100*	(5)
Greece	100*	(100)*	100*	(67)	100*	(33)	100*	(50)	100*	(83)*	100*	(33)
Hungary	100*	(87)*	100*	(67)	100*	(53)	100*	(47)	100*	(93)*	100*	(53)
India	100*	(73)	100*	(76)*	96*	(42)	96*	(35)	100*	(78)*	88*	(35)
Israel	100*	(93)*	100*	(73)	100*	(33)	100*	(40)	100*	(53)	80*	(13)
Italy	100*	(82)*	100*	(73)	100*	(46)	100*	(27)	91*	(36)	55	
Japan	100*	(76)*	100*	(74)	100*	(53)	96*	(33)	100*	(45)	88*	(20)
Norway	100*	(100)*	100*	(100)*	100*	(67)	100*	(50)	100*	(17)	50	(33)
Sweden	100*	(90)*	100*	(100)*	100*	(70)	100*	(35)	100*	(25)	65	(5)
Switzerland	100+	(100)*	100*	(80)*	100*	(80)	100*	(20)	100*	(40)	60	(20)
Turkey	100*	(80)*	100*	(50)	100*	(50)	67	(67)	100*	(100)*	100*	(50)
United Kingdom	100*	(100)*	100*	(97)*	100*	(71)	100*	(41)	100*	(7)	71	(3)
United States	100*	(97)*	100*	(90)*	100*	(64)	99*	(43)	97*	(19)	78*	(12)

Table 4A. Seven Goals of Counselling. Percent believing important[a] (Continued)

Country	Goal					
	Helping individuals/couples understand their options and the present state of medical knowledge so they can make informed decisions	Helping individuals/couples adjust to and cope with their genetic problems	The removal or lessening of patient guilt or anxiety	Helping individuals/couples achieve their parenting goals	The prevention of disease or abnormality	Improvement of the general health and vigor of the population
Total (95% Confidence limits)	99.8* (92)*	99.5* (87)*	99.7* (62)	98* (42) (96, 100)	97* (31) (95, 99)	74 (14) (69, 79)
Total, Excluding U.S. (95% Confidence limits)	99.7* (89)*	100* (85)*	99.7* (61)	98* (40) (96, 100)	98* (41) (96,100)	70 (15) (64, 76)
United States 1979[b](n=205) (95% Confidence limits)	Not Asked	100* (83)*	100* (75)	99* (44)	98.5* (52)	74 (11) (66, 82)

[a] In parentheses, percent believing the goal is "Absolutely Essential".
[b] Data from Sorenson et al. (1981)

Table 4B. Directive/Nondirective Approaches to Counselling

Country	Per Cent Believing Always/Sometimes Appropriate				
	Suggest that while you will not make decisions for patients you will support any they make	Tell patients that decisions, especially reproductive ones, are theirs alone and refuse to make any for them	Inform patients what most other people in their situation have done	Inform patients what you would do if you were in their situation	Advise patients what they ought to do
Australia	100*	92*	92*	8+	8+
Brazil	84*	94*	59	13+	16+
Canada	94*	94*	83*	45	9+
Denmark	100*	100*	62	31	8+
Federal Republic of Germany	91*	89*	48	30	9+
France	88*	73	31	38	25+
German Democratic Republic	90*	86*	52	10+	0+
Greece	86*	100*	57	29	14+
Hungary	80*	53	53	40	40
India	77*	89*	77*	58	56
Israel	93*	100*	67	0+	20+
Italy	91*	82*	46	9+	18+
Japan	90*	81*	58	45	27
Norway	83*	83*	50	50	0+
Sweden	90*	100*	47	11+	10+
Switzerland	100*	75*	100*	50	0+
Turkey	50	100*	75*	75*	50
United Kingdom	100*	94*	94*	30	0+
United States	98*	97*	69	19+	12+
Total (95 % Confidence limits)	94* (91, 97)	92* (89, 95)	66 (61, 71)	26 (22, 30)	15+ (12, 18)

Table 4B. Directive/Nondirective Approaches to Counselling (Continued)

Country	Per Cent Believing Always/Sometimes Appropriate				
	Suggest that while you will not make decisions for patients you will support any they make	Tell patients that decisions, especially reproductive ones, are theirs alone and refuse to make any for them	Inform patients what most other people in their situation have done	Inform patients what you would do if you were in their situation	Advise patients what they ought to do
Total, Excluding U.S.	90*	89*	64	33	18+
(95 % Confidence limits)	(86, 94)	(85, 93)	(58, 70)	(27, 39)	(14, 22)
United States 1979[a] (n=205)	93*	99*	66	18+	13+
(95 % Confidence limits)	(88, 98)	(97, 100)	(57, 75)	(12, 24)	(8, 18)

[a] Data from Sorenson et al. (1981)
* Strong (≥ 75 %) consensus of appropriateness
+ Strong (≥ 75 %) consensus of inappropriateness

would tell patients that decisions were the patient's alone. These are perhaps the supreme tests of nondirectiveness. Informing patients "what other people in their situation have done" is considered directive by some and nondirective by others – hence the back of consensus on this question. The last two approaches, telling patients "what you would do if you were in their situation" and advising patients "what they ought to do" are definitely directive, but 26 % and 15 %, respectively, considered these approaches appropriate.

Today's "nondirective" approaches to counselling actually have historical roots – now conveniently forgotten – in the American eugenics movement and are based on some eugenic assumptions. Sheldon Reed, who coined the term "genetic counselling" in 1947 (as more egalitarian than "genetic advice" or "genetic hygiene"), believed that, given adequate information, most patients would make "rational" decisions (e.g., not to have children if they were at high risk for serious mental disorders). According to Reed (1952, p.43):

"If our observation is generally correct, that people of normal mentality, who thoroughly understand the genetics of their problems, will behave in the way that seems correct to society as a whole, then an important corollary follows. It could be stated as a principle that the mentally sound will voluntarily carry out an eugenics program which is acceptable to society if counseling in genetics is available to them".

Furthermore, Reed believed that those who sought genetic counselling were so superior in intellect and character that they would on balance make a positive contribution to the gene pool even if their reproductive decisions were not always rational. In other words, genetic counselling could afford to be nondirective because:

1. most patients would make eugenically rational decisions on the basis of information provided, and
2. most patients who sought out the service were genetically superior in intelligence, moral character, and parenting abilities and therefore ought to have children anyway, even if some of their decisions were dysgenic for a single gene (Reed 1954).

As counselling evolved, it adopted a stated policy of nondirectiveness, best outlined in the 1974 *American Journal of Human Genetics* article on which our 12 questions in Tables 4A and 4B were based (Fraser 1974). Although the counselling profession forgot its eugenic roots, counsellors continued to report greater professional satisfaction in counselling well-educated, upper-middle-class people like themselves than in counselling those with less education (Wertz et al. 1988b). They communicated far better with well-educated people (Wertz et al. 1988a) than with racial minorities or poor people, half of whom did not show up for their appointments (Rapp 1988). As Reed predicted, counselling in the U.S. is disproportionately serving the well-educated and well-off. Geneticists in other nations in our survey – including some with national health insurance – expressed the same ethical concerns about class disparities.

Is counseling truly nondirective? This question is difficult to answer. Two of our case vignettes in the survey dealt with directiveness in counselling and also involved mental disorders. The first described a couple whose fetus was diagnosed with XYY and asked how the respondent would counsel prospective patients. Choices were:

1. Advise them to carry the pregnancy to term.
2. Advise them to have an abortion.
3. Refuse to make any suggestions about what they should do.
4. Provide complete information about the disorder, including prognosis, potential treatment and education, and possible problems the child may present for parents and society, taking care to present both sides of any topic that is controversial, and then let them make their own decision.
5. Give them the information described under reply choice 4 in an optimistic form so that they will favor carrying the pregnancy to term.
6. Give them the information under reply choice 4 but stress costs and burdens so that they will favor abortion without your having suggested it directly.
7. Give them information under reply choice 4 but also describe the emotional difficulties associated with terminating the pregnancy and then tell them to make their own decision.

Responses are described in Table 5. The columns headed "advise carry to term" and "advise abortion" include those who would give purposively slanted information. In all, 84 % would be nondirective, including 20 % who would include a discussion of the emotional difficulties associated with terminating the pregnancy. Women, who comprised 35 % of survey respondents, were 3.45 times more likely than men to be nondirective. In giving reasons for their choices, respondents demonstrated considerable variation in their perceptions of the seriousness of XYY. A total of 20 % considered XYY boys within the range of normal, despite the genetic diagnosis; others considered such boys a burden to society.

Our second nondirective counselling case involved tuberous sclerosis, a dominant disorder that is variably expressed. The question was phrased as follows:

"Evaluation of a child produces findings consistent with a diagnosis of tuberous sclerosis. On examining the parents, you find evidence that the father is carrying the tuberous sclerosis gene, even though he is of normal intelligence.

After a discussion of the risk of having another child with tuberous sclerosis who might be severely affected, the couple asks you whether recurrence of the disorder can be prevented. What course of action do you take with regard to telling the couple about EACH of the following options: taking their chances, artificial insemination by donor (A.I.D.), vasectomy, adoption, contraception".

The same question was repeated, with the mother as the carrier. Options were: taking their chances, contraception, tubal ligation, adoption, donor egg and IVF, surrogate motherhood.

Responses are described in Table 6. There was strong consensus ($\geq$ 75 % in 75 % of nations) about nondirective presentation of only one option, adoption. Geneticists in 14 nations would discuss this: exceptions were the German Democratic Republic and Hungary, where respondents said no babies were

Table 5. Counseling about Fetuses with XYY

Country	% Choosing Counseling Approach		
	XYY		
	Nondirective counseling	Advise carry to term[a]	Advise abortion[b]
Australia	92*	8	0
Brazil	84*	13	3
Canada	91*	9	0
Denmark	93*	7	0
Federal Republic of Germany	76*	24	0
France	35	65	0
German Democratic Republic	43	43	14
Greece	86*	14	
Hungary	33	60	7
India	77*	8	15
Israel	80*	20	0
Italy	82*	18	0
Japan	78*	14	8
Norway	83*	0	17
Sweden	90*	10	0
Switzerland	80*	20	0
Turkey	80*	0	20
United Kingdom	91*	9	0
United States	89*	9	2
Total (95 % confidence limits)	84* (80, 88)	14	2
Total, excluding U.S. (95 % confidence limits)	84* (75, 85)	17	3

[a] Includes giving optimistically slanted information
[b] Includes giving pessimistically slanted information telling parents that there is no major abnormality
* Strong consensus for nondirectiveness

available; India and Japan, where adoption is socially unacceptable to many people; and Turkey. Some newer options, IVF with donor egg and surrogate motherhood, were not available in most countries. In several countries (including Brazil and Hungary), A.I.D. had been illegal until recently.

In general, there was less consensus about presenting reproductive options than about prenatal counselling for XYY. It is difficult to "present" an option unasked, without leading a patient to think that you also recommend it. On the other hand, if you wait for patients to ask, many will not ask, either out of deference to the doctor or from lack of knowledge. If an option is *not* presented, many will think that you oppose it.

Although unbiased presentation in counselling is difficult, we believe that alternative methods of reproduction could help many families to solve their

Table 6. Presenting Reproductive Options. Percent of respondents who would present option, without giving advice, in cases where one parent carries tuberous sclerosis, a dominant disorder not diagnosable prenatally

Country	Option						
	Adoption[a]	Artificial Insemination-Donor	Taking Their Chances[a]	Contraception[a]	Sterilization[a]	Donor Egg, IVF	Surrogate Mother
Australia	100*	100*	64 (58)	75*	91* (75)*	92*	46
Brazil	78*	50	66	63	69	59	34
Canada	98*	98*	86*	82*	89*	59	37
Denmark	93*	80*	53	87*	67 (60)	80*	47
Federal Republic of Germany	85*	72	41	68	72	33	16$^+$
France	77* (71)	81*	71	53	53 (47)	44	0$^+$
German Democratic Republic	35 (55)	55	30 (80)*	25$^+$ (35)	30 (0)$^+$	25$^+$	5$^+$
Greece	80*	67	40 (33)	60 (50)	50 (17)$^+$	80*	25$^+$
Hungary	57 (62)	47	36 (43)	73 (31)	77* (39)	46	31
India	54	54	44	50 (41)	45	43	48
Israel	86*	100*	73 (69)	67 (77)*	67	80*	39
Italy	91*	80*	82*	82*	64 (55)	64	27
Japan	25$^+$ (20)*	26	58	39 (36)	19$^+$	21$^+$	10$^+$
Norway	83* (50)	83*	67	50 (67)	50	33	33
Sweden	86*	90*	68	85* (90)*	84*	55	30
Switzerland	80*	100*	60	60	60	80*	20$^+$
Turkey	60	60	80*	75* (60)	60	60	40
United Kingdom	93*	94*	94*	90*	90*	63	27
United States	95*	96*	88*	85*	84*	83*	67
Total	85* (84)*	83*	75*	74	74 (73)	66	46
(95% confidence limits)	(81, 89)	(79, 87)	(79, 80)	(69, 79)	(69, 79)	(61, 71)	(41, 51)
Total, excluding U.S.	76* (75)*	72	64	66	66 (64)	52	28
(95% confidence limits)	(70, 82)	(66, 78)	(58, 70)	(60, 72)	(60, 72)	(46, 58)	(23, 33)

[a] Questions asked for both male and female carriers. Percents are for female carriers: percents for male carriers are given in parentheses in countries where these differed from those for females.
* Strong consensus in favor of presenting
$^+$ Strong consensus against presenting

reproductive dilemmas and could also prevent children from suffering. As yet, these methods are rarely used in genetics. Ethical counselling should include presentation of all available reproductive options, if applicable to the situation. Donor eggs or surrogacy would be appropriate only for Mendelian dominant disorders where the woman was the carrier. For recessive disorders, A.I.D. would be the only ethical choice among the alternative methods of reproduction. Egg donation or surrogacy, because of the high physical and psychological costs to the donor, cannot be justified if A.I.D. is also a possible solution to the genetic problem. No alternative methods of reproduction should be presented as solutions for multifactorial disorders. Our scientific knowledge of the genetics of these disorders is too uncertain, and patients who resort to alternative methods of reproduction usually want "perfect children".

Geneticists may think of themselves as nondirective, but patients may nonetheless be influenced. In a 1977–79 survey in the U.S., 44 % of patients reported, six months after counselling, that the counselling had "influenced" their reproductive plans, but a comparison between the before-and-after counselling plans of the 44 % influenced and the 56 % not influenced showed no differences between the two groups (Wertz and Sorenson 1986). Those who said that they were influenced were the better-educated and may have used genetic information to bolster their pre-existing plans.

A more recent study or parents of children with cystic fibrosis (CF) suggests that patients' *perception* of the doctor's point of view can influence their decisions (Wertz et al., 1990b). Parents who perceived their CF doctors as supportive of selective abortion for CF were 1.86 times more likely to say they would have prenatal diagnosis for CF than were parents who perceived their doctors as not supportive of selective abortion. These CF doctors were not geneticists; we do not know whether they made any direct statements on the issue. Much doctor-patient communication is subtle, through gesture, demeanor and bearing, without open verbalization. It is the patient's *perception* of the doctor's view that counts, even if this perception is inaccurate.

In counselling, the interpretation of risk is crucial, especially for multifactoral disorders where numerical risks are uncertain. Patients generally tend to interpret a given numerical risk as lower than do geneticists. Geneticists in the Sorenson study interpreted risks of 7–19 % as moderate while patients interpreted risks up to 15 % as low; geneticists interpreted risks of 20–24 % as high and risks of 25 % or more as very high, while patients interpreted risks up to 50 % as moderate (Wertz et al. 1986). Geneticists should remember that a risk finding that looks impressive in the laboratory may have little influence on patients, especially in multifactorial disorders where genetic studies indicate risks under 15 %.

Counselling should be conducted in a manner accessible and understandable to persons of all social groups. Although we agree that, in general, nondirectiveness is a laudable intent, we believe that there may be some situations calling for directive advice (see *Disclosure* above) and some parental choices that geneticists need not "support". It is time for the profession to reconsider unqualified nondirectiveness.

Screening

Although population screening for most mental disorders is not feasible at this time, it is not premature to consider general ethical guidelines for screening in order to prevent coercion or stigmatization. At present geneticists are divided about the primary goals of population screening in general and the age groups that should be screened. Our survey question about a carrier test for CF – in 1985, a hypothetical test – revealed consensus that screening be voluntary, but no consensus about the age group to be screened first (Table 7). Mandatory screening of newborns has generally been reserved for conditions, such as PKU, where early treatment is available and offers significant benefit to the child (President's Commission 1983, p.58). More recently, some have suggested "routine" newborn screening for conditions, such as Duchenne muscular dystrophy, where early diagnosis provides no significant benefit to the child. The purpose of routine newborn screening for Duchenne is to identify carrier couples and counsel them before the wife becomes pregnant again (Knoppers and Laberge 1990). Although ethicists on the President's Commission (1983)

Table 7. Group to receive mass carrier screening for cystic fibrosis

Country	Percent responding[a]				
	Newborns, by law	Newborns, by consent	Children < 18, by consent	Adults > 18, by consent	Other ages, by law
Australia	0	59	8	25	8
Brazil	25	16	0	53	6
Canada	18	27	11	42	2
Denmark	27	20	20	33	0
Federal Republic of Germany	11	30	9	47	3
France	35	12	6	35	12
German Democratic Republic	67	0	5	10	18
Greece	0	0	0	86*	14
Hungary	67	7	0	7	19
India	39	31	4	23	3
Israel	20	7	7	53	13
Italy	18	0	27	46	9
Japan	0	51	15	30	4
Norway	33	33	0	34	0
Sweden	5	29	0	66	0
Switzerland	0	80*	0	20	0
Turkey	80*	0	0	0	20
United Kingdom	0	30	27	37	6
United States	23	22	17	35	3
Total	20	25	13	37	5
Total, excluding U.S.	18	28	10	39	5

[a] Indicates strong (≥ 75 %) consensus.

opposed mandatory or routine newborn screening except for the direct benefit of the newborn, arguing that mandatory screening for purposes of carrier identification was an invasion of privacy, there is a growing movement in the United States to use routine newborn screening primarily for identification of carrier parents.

In giving reasons for their answers to our CF screening questions, respondents were divided about whether screening should be used to identify and treat newborns, to identify carrier parents before the next pregnancy, to identify teenage carriers before marriage, or to identify adult carriers who were considering children. Utilitarian issues also played a role; many believed that newborns were the easiest group to screen, but recognized that *carriers* identified as newborns might never learn their carrier status or might forget it.

There was a consensus (75%) that screening, at whatever age, be voluntary rather than mandatory. This consensus may change in the light of increasing evidence that early treatment benefits newborns with CF. The screening situation is further confused by the fact that, in many jurisdictions, there is a sliding scale between "mandatory" and "voluntary". Many legally mandated newborn screening programs allow parents to refuse screening ("informed refusal"), though few parents do so. Many voluntary, but routine, screening programs that require parents to sign informed consent forms do not explain the meaning of these forms; parents sign because they think that refusal to do so will prejudice their child's care.

We suggest that *all* genetic screening be strictly voluntary, except for newborns when early treatment is available and beneficial. Mandatory or routine screening should not be performed for purposes of carrier identification. Genetic testing should, however, be offered on a voluntary basis and should be financially accessible to all. Testing should be accompanied by unbiased information about the disorder in question and by counselling for those who test positive.

Prenatal Diagnosis

It is somewhat premature to discuss prenatal diagnosis for mental disorders, with the obvious exceptions of trisomies and single-gene disorders. Nevertheless, in the future some may suggest that prenatal diagnosis also be applied to multifactorial disorders. Such applications are ethically and scientifically questionable, because they are likely to have low predictive value. Applying prenatal diagnosis to common mental disorders such as schizophrenia- if this becomes possible- could place an unpleasant burden of choice on parents. Once prenatal diagnosis is used on a large scale for a particular disorder, carrying an affected fetus to term may be perceived as a voluntary act for which parents are personally responsible, rather than an act of nature. Affective mental disorders are particularly costly to treat, and many private insurance policies in the U.S. already exclude all but the first 30 days of hospitalization. Insurance companies may try to exclude affected persons whose mental disorders or predispositions to mental disorders were diagnosed prenatally, arguing that the parents could have prevented the birth. In countries whose governments underwrite heath care, there may be political pressure to exclude coverage for such individuals. This would be unjust and would rob many prospective parents of free choice.

Most geneticists (83 %) in our survey believed that prenatal diagnosis should be offered without preconditions to parents who refused abortion for serious fetal defect (Table 8, column 1), but this included 34 % who thought that parents might change their minds after a positive diagnosis. There were exceptions: in India, Norway, and Turkey prenatal diagnostic capacities are limited, and many geneticists felt that using prenatal diagnosis to prepare for the birth of an affected child was a waste of scarce or costly resources. In all, women were 3.96 times more likely than men to say that they would perform prenatal diagnosis.

Prenatal diagnosis has two avenues of application in psychiatry: to alleviate maternal anxiety and to diagnose mental disorders in the fetus. A substantial number (63 %) of geneticists would perform prenatal diagnosis for an anxious woman of 25 with no family history of genetic disorders and no personal exposure to toxic substances (Table 8, column 2). The case vignette described no psychiatric or mental condition in the mother, only that "she appears very anxious about the normalcy of the fetus".

Geneticists who would perform prenatal diagnosis for this woman gave, as reasons, respect for patient autonomy (56 %) and prevention of harm to the patient by alleviating anxiety (46 %). Very few would refer the anxious patient to a psychiatrist for diagnosis or therapy. In Denmark, Sweden, and Switzerland, anxiety is already considered a valid medical indication for prenatal diagnosis, and accounts for 7–11 % of procedures (Wertz and Fletcher 1989).

Prenatal diagnosis on grounds of maternal anxiety is essentially prenatal diagnosis on demand, especially if there is no psychiatric confirmation of excessive or morbid anxiety. Most pregnant women are anxious, and anxiety can be used to conceal another reason for seeking prenatal diagnosis, namely sex selection. We have urged that prenatal diagnosis be limited to cases where there is an excess risk of fetal defect, though it could also be applied on the mother's behalf in rare instances where psychiatric consultants recommended it to alleviate morbid anxiety. Large-scale application of prenatal diagnosis to the normal anxiety of pregnancy not only increases risks to the fetus, but promotes class disparity in the use of genetic services, because most such requests will probably come from educated people with higher incomes.

Prenatal diagnosis may have less impact than geneticists anticipate, especially for late-onset or behavioral disorders. Preliminary data from a study of 271 parents of children with CF (Wertz et al. 1990) suggest that although the majority would abort for severe mental retardation, few would abort for a "severe painful, incurable disorder starting at age 40", a "severe, painful disorder starting at age 60", or "strong genetic susceptibility to alcoholism" (Fig. 1, black bars). Socio-demographically, these parents represent a cross-section of the white population in six states in the northeastern U.S. Their views on abortion for maternal or family situations (not shown in table) closely approximate the results of national public opinion surveys. Thus there is reason to believe that their views on abortion for fetal conditions may be representative

Table 8. Indications for prenatal diagnosis. Percent of respondents who would perform prenatal diagnosis or refer to someone who would perform it

Country	Patients who refuse abortion		Maternal anxiety only	
	Perform	Refer	Perform	Refer
Australia	92*		50	17
Brazil	78*	3	14+	30
Canada	87*	2	55	15
Denmark	87*		87*	
Federal Republic of Germany	85*		73	7
France	56		38	19
German Democratic Republic	54		5+	
Greece	57		43	
Hungary	47		20+	
India	56	7	41	15
Israel	93*		80*	
Italy	73	18	73	9
Japan	44	4	16+	2
Norway	50		50	
Sweden	91*		91*	
Switzerland	100*		100*	
Turkey	60		20+	20
United Kingdom	91*		79*	9
United States	96*		78*	11
Total	83*	2	63	10
(95 % confidence limits)	(79, 87)		(57, 69	
Total, excluding U.S.	74	2	53	8
(95 % confidence limits)	(68, 80)		(47, 59)	

* Strong consensus (≥ 75 %) in favor of performing
+ Strong consensus (≥ 75 %) against performing

of the wider U.S. population. If this is so, future use of prenatal diagnosis and selective abortion may be comparatively rare, except for the most devastating mental and physical disorders that appear in early childhood.

The uptake of prenatal diagnosis for mental disorders will depend in large part on public perception of the disorder in question. It is of the utmost importance that prenatal testing be accompanied by a full, fair, and unbiased description of the disorder, including the full range of severity, the effects on family life, and how an affected person may fare as an adult. Video-tapes could be helpful. Fair presentation of each disorder, without hidden agendas, is absolutely crucial to ethical testing. Any attempts to bias educational programs for purposes of reducing the costs of caring for affected children or for eugenic purposes could

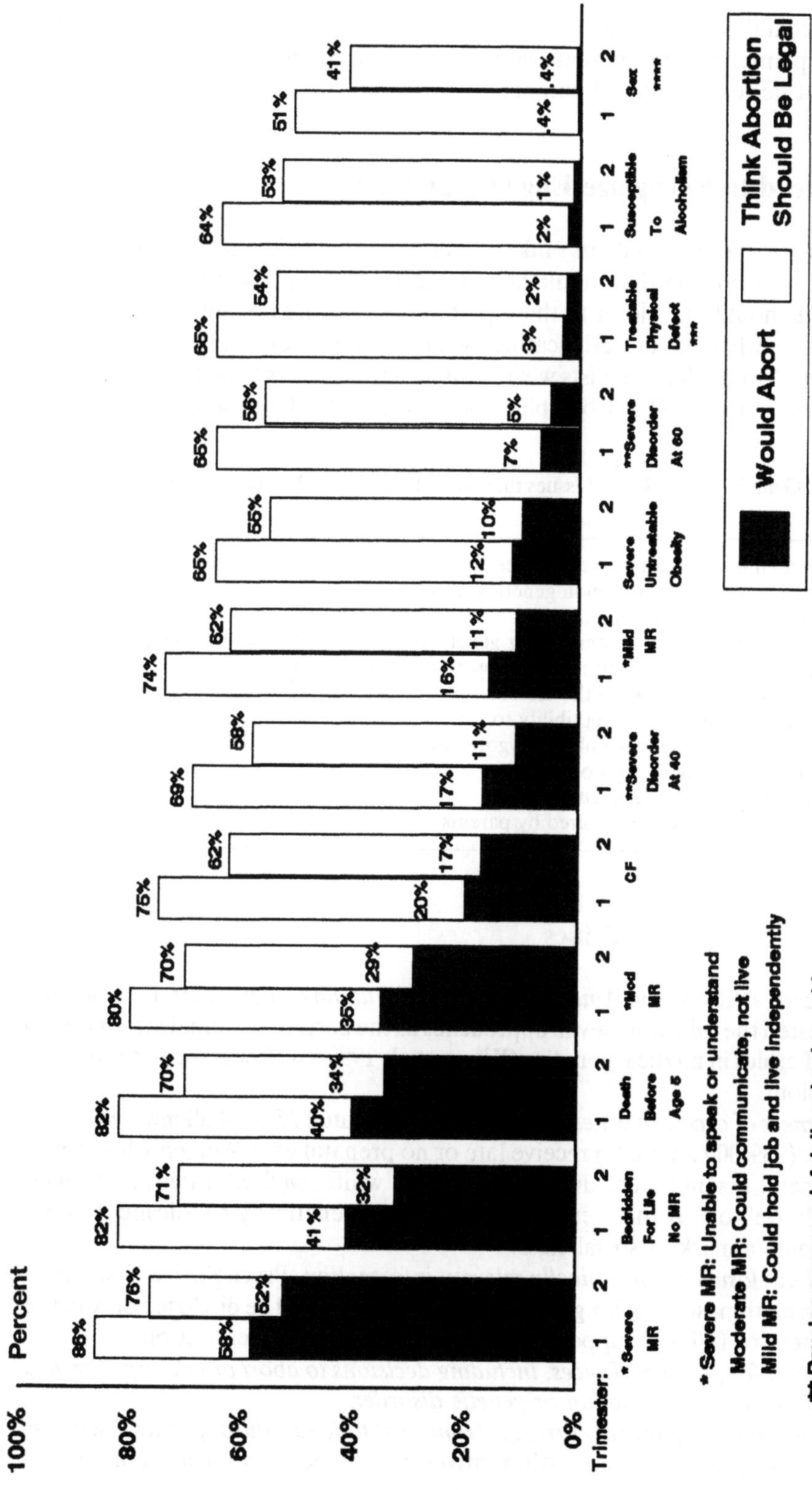

Fig. 1. Attitudes Toward Abortion Among Parents of Children with Cystic Fibrosis (N=271)
Fetal Characteristics

From: Wertz et al. 1990b

backfire and discredit modern genetics. On the other hand, the optimistic picture painted by some parents who have learned to cope with affected children may not be realistic for all parents.

The Need to Recognize Eugenic Concerns

Although survey respondents ranked "long-range eugenic concerns" ninth out of a list of ten priorities that should be of ethical concern (Table 9), we believe that eugenics should rate closer to the top of the list. We are not advocating eugenic programs; what we are advocating is an awareness of "hidden agendas" in genetics. Although geneticists wish to divorce themselves from the sorry history of eugenics, they will not be able to accomplish this by avoiding the issue.

Table 9. Geneticists' Ranking of Issues that Should Be of Most Ethical Concern in Next 10–15 Years

1.	Increased demand for genetic services
2.	Carrier screening for common genetic disorders
3.	Allocation of limited resources
4.	Development of new treatments for genetic disorders, including treatment in utero, organ transplantation, and molecular genetic manipulation
5.	Environmental damage to the unborn
6.	Screening for genetic susceptibility to cancer, heart disease
7.	Research on the human embryo, zygote, and fetus
8.	Genetic screening in the workplace
9.	Long-range eugenic concerns
10.	Sex preselection for sex desired by parents

Need for a Code of Ethics

In a recent article in the *American Journal of Human Genetics,* Dr. Fletcher and I suggested the following seven approaches as the basis for a formal international code of ethics in medical genetics (Wertz et al. 1990a). Now is the time to begin discussion.

1. *Fairness of access to genetic services.* An estimated 25 % of all mothers in the U.S. (939,000 annually) receive late or no prenatal care. Women who receive prenatal diagnosis are disproportionately white, well educated, and financially well-off. It this trend continues, being genetically handicapped could become a mark of social class.
2. *Full disclosure of all clinically relevant information.* Psychologically sensitive information, such as XY genotype in a female, should be disclosed only in the context of full and supportive counseling and patient education.
3. *Respect for parental choices, including decisions to abort or to carry to term a fetus with a malformation or genetic disorder.*
4. *Protection of patients' privacy from institutional third parties,* such as employers and insurers. Alternatives would be a) to permit access to

information but to have extensive legal protection for individuals that would prevent discrimination, or b) government underwriting of health insurance for those at genetic risk.

5. *Use of prenatal diagnosis only to give parents information about the health of the fetus.* Any other use, such as for sex selection (except for X-linked disease), should be avoided.
6. *Voluntary, not mandatory, screening – Except for newborns when early treatment is available.* The primary purpose of screening should be to help the newborn; carrier detection is secondary.
7. A need for further study and discussion of the complex issue of patient confidentiality in cases where there is high risk of serious harm to relatives at genetic risk.

References

Billings PR (1990) Brewing genes and behavior: the potential consequences of genetic screening for alcoholism. In: The molecular biology and genetics of alcoholism. Cold Spring Harbor, Cold Spring Harbor (Banbury conference, vol 33)

Chorover SL (1979) From genesis to genocide: The meaning of human nature and the power of behavior control. MIT Press, Boston, pp 57–76

Czeizel A (1988) The right to be born healthy: the ethical problems of human genetics in Hungary. Liss, New York

Davenport CB (1919) Naval officers: Their heredity and development. Carnegie Inst Washington Publ 259

DeLee JB (1920) The prophylactic forceps operation. Am J Obstet Gynecol 1:34–44

Fink AE (1938) Causes of crime: biological theories in the United States, 1800–1915. University of Pennsylvania Press, Philadelphia

Fraser FC (1974) Genetic counseling. Am J Hum Genet 26:636–659

Holtzman NA (1989) Proceed with caution: predicting genetic risks in the recombinant DNA era. Johns Hopkins University Press, Baltimore

Kevles D (1985) In the name of eugenics: genetics and the uses of human heredity. University of California Press, Berkeley

Knoppers BM, Laberge CM (1990) Genetic screening: from newborns to DNA typing. Elsevier, Amsterdam, pp 111–149

Paul D (1991) Social origins of clinical genetics. In: Juengst E (ed) Philosophical and historical problems in medical genetics. Reidel, Amsterdam (in press)

President's Commission for the Study of Ethical Problems in Medicine and Biomedical and Behavioral Research (1983) Screening and counseling for genetic conditions. US Government Printing Office, Washington

Rafter NF (1989) White trash: the eugenic family studies. Northeastern University Press, Boston

Rapp R (1988) Chromosomes and communication: the discourse of genetic counseling. Med Anthropol Q 2:143–157

Reed SC (1952) Heredity counseling and research. Eugen News 37:41–46

Reed SC (1954) Heredity counseling. Eugen Q 1:47–51

Sorenson JR, Swazey JP, Scotch NA (1981) Reproductive pasts, reproductive futures: genetic counselling and its effectiveness. Liss, New York

Wertz DC (1990) Ethical issues in the application of knowledge from molecular genetics to mental disorders. In: Bulyzhenkov V, Christen Y, Prilipko L (eds) Genetic approaches in the prevention of mental disorders. Proceedings of the joint meeting organized by the World Health Organization and the Fondation Ipsen, Paris, May 29–30, 1989. Springer, Berlin Heidelberg New York pp 92–106

Wertz DC, Fletcher JC (1988) Attitudes of genetic counselors: a multi-national survey. Am J Hum Genet 42:592–600
Wertz DC, Fletcher JC (1989) Ethics and human genetics: a cross-cultural perspective. Springer, Berlin Heidelberg New York
Wertz DC, Sorenson JR (1986) Client reactions to genetic counseling: self-reports of influence. Clin Genet 30:494–502
Wertz DC, Sorenson JR, Heeren TC (1986) Clients' interpretations of risks provided in genetic counseling. Am J Hum Genet 39: 253–264
Wertz DC, Sorenson JR, Heeren TC (1988a) Communication in health professional-lay encounters. In: Ruben BD (ed) Information and behavior 2. Transaction, New Brunswick, pp 329–432
Wertz DC, Sorenson JR, Heeren TC (1988b) 'Can't get no (dis)satisfaction': professional satisfaction with professional-client encounters. Work Occup 15:36–54
Wertz DC, Fletcher JC, Mulvihill JJ (1990a) Medical geneticists confront ethical dilemmas: cross-cultural comparisons in 18 nations. Am J Hum Genet 46:1200–1213
Wertz DC, Rosenfield JM, Janes SR, Erbe RW (1990b) Attitudes toward termination of pregnancy among parents of children with cystic fibrosis. Presented at American Society of Human Genetics, Cincinnati, Oct. 19, 1990.
Wertz RW, Wertz DC (1989) Lying-in: a history of childbirth in America. Yale University Press, New Haven

The Use of Prenatal Diagnosis for Psychiatric Diseases

J.-F. Mattei

Prenatal diagnosis classically concerns the screening during pregnancy for diseases characterized by an organic or mental expression whose mode of transmission and genetic markers are known and specific, independently of the surrounding environment.

In psychiatric pathology, polyfactorial heredity is involved, implicating susceptibility genes whose expression depends more or less on the surrounding environment.

Recent developments in molecular biology now allow diagnosis of genetic prenatal anomalies whose pathological character, risk of appearance – which may be deferred or uncertain – and actual severity are very difficult to determine a priori.

All of these problems pose the question of the ethical considerations which are implicated in the development of biological and medical techniques.

The psychiatrist and the geneticist have long been set at odds, often in a manner resembling caricature. One dealt with mental pathologies and behavioral disorders linked to the environment; the other was oriented towards the study of genetic material which is known to contain the genes destined to define the biological originality of the individual.

In fact, this opposition between psychiatry and genetics has served to artificially fuel the eternal philosophical debate between nature and nurture, but the boundaries are now becoming blurred. What was at first a discreet flirtation has become an open relationship, and today's topic shows that the union has been consummated.

The Problem of Definitions

Normal and abnormal behavior

One must first of all define psychiatric disease. What is normal behavior? What is abnormal behavior? What is the boundary? These questions are highly problematical, and subjectivity plays an important role, as does the reference environment, inasmuch as behavior is differentially considered depending on the conditions of daily life. Nevertheless, it is true that certain psychiatric

Srám, Bulyzhenkov, Prilipko, Christen (Eds.)
Ethical Issues of Molecular Genetics in Psychiatry

disorders may be defined precisely, for example, schizophrenia, manic-depression and autism.

I have deliberately mixed elements which probably cannot be compared because this brings up the question of etiology.

Diseases and Syndromes

The question of etiology must be addressed before one deals with the possibility of prenatal diagnosis. Over the years psychiatry, one of the oldest disciplines, has founded itself on clinical criteria. Perhaps more than in other disciplines, psychiatrists refer to a disease or a syndrome without bearing in mind the fundamental difference which separates them, particularly as concerns the present topic. Even if, for a disease, the same clinical picture always corresponds to a single cause, we all know that a syndrome is a clinical expression which is common to different causes which generally have nothing to do with one another.

As geneticists, our first problem is to recognize the genetic cause behind a clinical picture which often appears in disparate conditions. One could choose many examples, all significant, but I would like to focus on autism. Indeed our team has acquired extensive experience in this subject through studies of mental retardation linked to fragile X (Mattei et al. 1981; Nguyen et al. 1989).

For a long time it was thought that autism in childhood was due to a lack of maternal affection. I recently reread the views of a well-known psychiatrist who maintained that nothing could be done for an autistic child if its mother would not first agree to be psychoanalyzed. For a long time people opposed psychiatry, which was linked to the surrounding environment, and genetics, which was linked to the transmission of hereditary traits. Yet we were contacted by families with children who were considered autistic. Clinically speaking, all of the symptoms were clear: isolation, introversion, refusal to communicate, no eye contact, absence of language, behavioral disorders, etc. The parents had first consulted a child psychiatrist about the child's language retardation and unstable personality when the child was 4. Autism was the most fequent diagnosis. Discreet facial dysmorphy and the incidence of recurrence in the boys in the family were also reported.

A study of the genealogical tree indicated the possibility of an X-linked recessive genetic transmission. We were most surprised to observe on karyotyping that the X chromosome was fragile on the distal part of the long arm. I shall not discuss here all the questions raised by this disease, which remains one of the most mysterious in genetic pathology, because of the Mendelian transmission of an illness resulting from a chromosomal anomaly and the impossibility, to date, of cloning, the gene (if it exists), despite the fact that research on the disease has been carried out since 1983 about as long as reseach on Duchenne muscular dystrophy (Camerino et al. 1983). Since then many other genes have been localized, identified and isolated using similar techniques, yet we and numerous other groups all over the world have failed to pierce the mystery of this fragile site.

Concerning autism, we now have proof that it is a syndrome and not a disease, that genetic factors play a considerable role, as indicated by the families studied, by studies on twins and by the higher incidence in boys (4 boys to 1 girl on average; (Jorde et al. 1990).

The following figures may be proposed: 10 % of autistic boys are fragile site carriers, 15 % of fragile X boys display autistic behavior.

This example points out the difficulties to be encountered; one can in no case launch oneself into genetic prenatal diagnosis without first of all having established the genetic cause of the disease in the family concerned.

This brings us back to current data on known and localized genes which clearly play a determining role in psychiatric pathology.

Direct and Indirect Genetic Approaches

Several theoretical situations must be distinguished. Either the gene is localized by linkage methods and the diagnosis is established by indirect approaches having a considerable risk of error (Lathrop et al. 1985), or the gene is known and identified and then direct diagnosis may be considered as being certain.

I have no examples of the second possibility; most of the genetic data on psychiatry result from linkage analysis, the first situation. It is by using this means that a schizophrenic gene was localized to the long arm of chromosome 5 in 5q11–q13.3 (Sherrington et al. 1988; McGillivray et al. 1990) and that manic-depression psychosis was linked to genes on the short arm of chromosome 11 and chromosome X. It should be noted, however, that biologists are not entirely convinced. In a "News and Views" article published in 1989 in *Nature,* Robertson stated that there were no thoroughly convincing data of genetic linkage between a psychiatric disorder and a given locus. He was contradicted by others who offered arguments to explain discrepancies in multiple genes, incomplete penetrance, heterogeneous causes and numerous other imperfections in pedigree analyses (Kennedy et al. 1988; Detera-Wadleigh et al. 1989; Byerley 1989; Edwards 1990; Gurling 1990). We must still agree on the localizations of Rett syndrome (Journel et al. 1990), Huntington chorea (Gusella et al. 1983), Alzheimer's disease (St. George-Hyslop et al. 1987), Tourette's syndrome (Merz 1988) or other neuropsychiatric disorders. This underlines the haziness of both clinical and genetic concepts in this difficult chapter of human pathology.

The Rationale Behind Prenatal Diagnosis

My presentation so far could be considered to be composed only of questions, uncertainties and hesitations. This was done deliberately to show how much geneticists are aware that genetics does not sum up science. To progress and serve, genetics needs to be elaborated in a logical and rigorous framework if it is not to lead to erroneous lines of conduct which would be all the more perverse as they would appear beyond any scientific reproach.

From Hazy Concepts to Rigorous Techniques

However, opposed to these uncertainties are our methods. These are codified, mastered and reassuring in their rigorous scientific protocols. We obviously know how to collect amniotic fluid after 16 weeks of pregnancy. We know how to obtain chorionic villi after 9 weeks and then use these samples to study the foetal chromosomes, analyse DNA with molecular biology techniques and assay various chemical compounds. One could even forget the haziness of the concepts behind the screen formed by these technical certainties.

Likewise, results will be interpreted as being normal or pathological because, even if the biologist is also hesitant in many situations, the couples and their physicians will react differently. Either the result is clearly normal or it is, whatever the risk of an anomaly, considered abnormal, with all the consequences which we can imagine. Termination of the pregnancy is nearly always requested.

If I insist on this point, it is because accepting the principle of prenatal diagnosis means implicitly accepting the logical consequences of such a step, despite inner doubts.

The Demand for Quality of the Ideal Child

Couples today have trouble understanding the uncertainties of science and medicine because they place increasingly higher quality demands on their offspring. Parents admit that there are things we do not know. They do not understand that we might have doubts. And they do not accept risk. One may cite the example of a chromosomic anomaly, the double Y chromosome. It is representative of how couples react (Mattei and Gamerre 1989).

While doing prenatal diagnosis for Down's syndrome because of maternal age, we sometimes detect another anomaly, for example, a double Y chromosome. As we inform the parents of this anomaly, it is quite difficult to explain, and have them accept all at once, the frequency of the anomaly, the fact that behavior may not be affected and that psychopathic disorders vary widely in their expression. Besides our information, they seek more information from more or less well-informed people and from mass readership medical books which often highlight the sensational aspects. They nearly always come back a few days later requesting termination of pregnancy. I am uncertain whether this path represents real progress for man.

Starting with a diagnosis for Down's syndrome, which is a sure diagnosis of severe and definitive mental retardation, we are now faced diagnoses whose risks are evaluated statistically and for which behavioral disorders vary considerably. We are engaged in a decision-making process whose logic, it seems to me, escapes the human condition itself. After having progressively developed the idea that now a child is more an object that we have the right to than a being whose presence is a gift ("to be or not to be," that is the question asked through medically assisted procreation), we are now asking "to be normal or not to be." It is evident that, by choosing to eliminate those children who may deviate from behavioral norms, we are heading towards choosing an a priori defined behavior. The question then becomes: "to be in conformity or not to be," which leads to the concept of the perfect child, the ideal child, one corresponding to the parents' imagination.

From Psychiatry to Genetics: To Be Normal or Not to Be

There are two other important aspects in prenatal diagnosis. The first is the relationship between psychiatry and the similitude of behaviors and decisions. The first responsibility of psychiatry is to define abnormal behavior, a heavy task. Its second responsibility is just as difficult: To determine the point at which an anomaly is so unbearable that it justifies internment, i.e., loss of liberty.

Genetics adopts a similar approach. Geneticists have the responsibility of defining what a normal human being is before establishing the gravity of the anomaly which would justify the loss of life. One cannot avoid mentioning this when dealing with prenatal diagnosis of psychiatric disease, which involves the rationale behind choosing to intern a child in death if the genetic patrimony he has been bequeathed should harbor the potential of a fate different from that of the majority, and impossible for the majority to understand. Yet, I am not sure that physicians have the vocation of judges.

Predicting Life and Changing Fate

An other aspect of prenatal diagnosis is equally important. The potential of having a different fate evokes two questions:

1. will the detected pathological gene necessarily be expressed? and
2. if so, during which period of life will it be expressed?

I would simply point out that we have reached a new level, that of predictive medicine. It is a fascinating and frightening field because, with the techniques of molecular biology, we will one day be able to detect strengths and weaknesses of individuals even though they are buried deeply within the nucleus of their cells. We shall soon, (and this is already being done) foretell what fate holds in store for us before life even starts, as if we are the fortune-tellers of modern times. Southern blots now replace the crystal ball.

These questions are directly linked to prenatal diagnosis since, if it becomes part and parcel of daily life, the choices and their consequences will be terrifying. If we consider fate to be inexorable, we will have to choose between terminating life or accepting it as it is: terminating the life of a being who is destined to be "too different" from the others in order to keep him from suffering from this difference. I am obviously not speaking of the embarrassment this different-life would cause society. No one thinks of his own comfort, only the suffering of others preoccupies us; we all know this in our good faith, which is equaled only by our dishonesty. What of accepting life as it is after prenatal diagnosis has indicated the possibility of a psychiatric disorder? If one pushes this thinking further in its reasoning and analyzes its consequences, there invariably appears an inevitable perversity. Of course, knowing about a risk may lead us to take preventive measures to try and modify the genetic fate as it is known, relying on the environment to play a considerable part. Nevertheless this gives society a role which is contrary to the one it should play.

Whereas society should seek to reduce inequalities between men, it would find itself ratifying these inequalities thanks to predictive medicine and, under the pretext that of protecting the whole, society would classify and intern such and such a person in such and such a category depending on his risk. This constitutes one of the most difficult issues among those we shall have to address.

Conclusion

I chose not to enter into precise details on indications. Each of us knows that sometimes decisions are imposed on us by particular circumstances and that there is no worse attitude than to shut oneself into simplistic and stereotyped positions which turn a deaf ear to distress. We must deal with our inner contradictions. I have chosen to present the general lines of reasoning brought up by prenatal diagnosis of psychiatric diseases. Evidently, these questions cannot but lead us on to the meaning of life, death and suffering and the very notions of liberty and fate.

To conclude, we know that the road ahead is long. Quite simply, today, given the present state of our scientific and ethical knowledge, I am far from convinced that prenatal diagnosis of psychiatric disease constitutes a desirable approach or that it represents true progress of science in the service of humankind.

References

Byerley WF (1989) Genetics linkage revisited. Nature 340:340–341

Camerino G, Mattei MG, Mattei JF, Jaye M, Mandel JL (1983) Close linkage of fragile X-mental retardation syndrome to hemophilie B and transmission through a normal male. Nature 306:701–704

Detera-Wadleigh SD, Goldin LR, Sherrington R, Encio I, de Miguel C, Berrettini W, Gurling H, Gershon ES (1989) Exclusion of linkage to 5q11–13 infamilies with schizophrenia and other psychiatric disorders. Nature 340:391–393

Edwards JH (1990) Genetic linkage and psychiatric disease (scientific correspondence). Nature 344:298–299

Gurling H (1990) Genetic linkage and psychiatric disease (scientific correspondence). Nature 344:298

Gusella JF, Wexler NS, Conneally PM, Naylor SL, Anderson MA, Tanzi RE, Watkins PC (1983) A polymorphic DNA marker genetically linked to Huntington's disease. Nature 306:234–238

Jorde LB, Mason-Brothers A, Waldmann R, Ritvo ER, Freeman BJ, Pingree C, McMahon WM, Petersen B, Jenson WR, Mo A (1990) The UCLA-University of Utah epidemiologic survey of autism: genealogical analysis of familia aggregation. Am J Med Genet 36:85–88

Journel H, Melki J, Turleau C, Munnich A, de Grouchy J (1990) Rett phenotype with X/autosome translocation: possible mapping to the short arm of chromosome X. Am J Med Genet 35:142–147

Kennedy JL, Giuffra LA, Moises HW, Cavalli-Sforza LL, Pakstis AJ, Kidd JR, Castiglione CM, Sjogren B, Wetterberg L, Kidd KK (1988) Evidence against linkage of schizophrenia to markers on chromosome 5 in a northern Swedish predigree. Nature 336:167–170

Lathorop GM, Lalouel JM, Julier C, Ott J (1985) Multilocus linkage analysis in humans: detection of linkage and estimation of recombination. Am J Hum Genet 37:482–498

Mattei JF, Gamerre M (1989) Ethics and medical genetics in France. In: Wertz DC, Fletcher JC (eds) Ethics and human Genetics. Springer, Berlin Heidelberg New York pp 176–193

Mattei JF, Mattei MG, Aumeras C, Auger M, Giraud F (1981) X-linked mental retardation with the fragile X. A study of 15 families. Hum Genet 59:281–289

McGillivray BC, Bassett AS, Langlois S, Pantzar T, Wood S (1990) Familial 5q11.2->q13.3 segmental duplication cosegregating with multiple anomalies, including schizophrenia. Am J Med Genet 35:10–13

Merz B (1988) Tourette's syndrome hypothesis creates controversy among geneticists. JAMA 260:2619

Nguyen C, Poutska AM, Djabali M, Roux D, Mattei JF, Lehrach H, Jordan BR (1989) Large-scale mapping and chromosome jumping in the q27 region of the human X chromosome. Genomics 5:298–303

Robertson M (1989) News and views. Nature 342:222

Sherrington R, Brynjolfson J, Petursson H, Potter M, Dudleston K, Barraclough B, Wasmuth J, Dobbs M, Gurling H (1987) Localization of a susceptibility locus for schizophrenia on chromosome 5. Naure 336:164–167

St George-Hyslop PM, Tanzi RE, Polinsky RJ, Haines JL, Nee L, Watkins PC, Myers RH, Feldman RG, Pollen D, Drachman D, Growdon J, Bruni A, Foncin JF, Salmon D, Frommelt P, Amaducci L, Sorbi S, Piacenti S, Stewart GD, Hobbs WJ, Conneally PM, Gusella JF (1987) The genetic defect causing familial Alzheimer's disease maps on chromosome 21. Science 235:885–890

New Ethical Problems Related to Environmental Pollution and Behavioral Changes in Human Population

R.J. Šrám

In our environment there are many factors which can be harmful to our genetic material. Mutagens can induce not only mutations in gametes, which increase spontaneous abortions, congenital anomalies (CGA) or cancer, but also mutations in somatic cells, which are responsible in cancer or aging. CGA are believed to be one of the best mutation endpoints for genetic monitoring in regions polluted by environmental mutagens. CGA can be understood as an endpoint which is determined by the heritage of both parents and the effect of mutagens on the gametes of both parents before the conception and on the foetus during the intrauteral development (which is simultaneously teratogenic). This means that the pregnancy outcome is determined by the parents' genetic predisposition to diseases, environmental pollution, occupational exposure and lifestyle (smoking, alcohol, diet, drug abuse) before the conception and, during the course of the pregnancy, also by the mother's infections, drugs taken and occupational exposure (Bochkov and Šrám 1989).

Environmental Problems in Northern Bohemia

To verify this idea, we studied the frequency of CGA in Northern Bohemia, which is believed to be one of the most polluted regions in Europe. This region is highly polluted by sulphur dioxide (SO_2), nitrogen oxides (NO_x), polycyclic aromatic hydrocarbons (PAH) and heavy metals from industrial sources due to geographic morphology and frequency of inversion situations. For these reasons we analysed pregnancy outcomes and their relationship to various factors in three districts of Northern Bohemia with specific types of environmental pollution:

1. the district of Ustí n.L. (district A), characterized by the chlorine chemical industry, with pollution from brown coal power stations and brown coal strip mines;
2. the district of Teplice (district B), polluted by brown coal power stations and brown coal strip mines; and,
3. the district of Jablonec n.N. (district C), the so-called "control" district, which is partially polluted by brown coal power stations in Poland and Germany (Fig. 1).

Srám, Bulyzhenkov, Prilipko, Christen (Eds.)
Ethical Issues of Molecular Genetics in Psychiatry

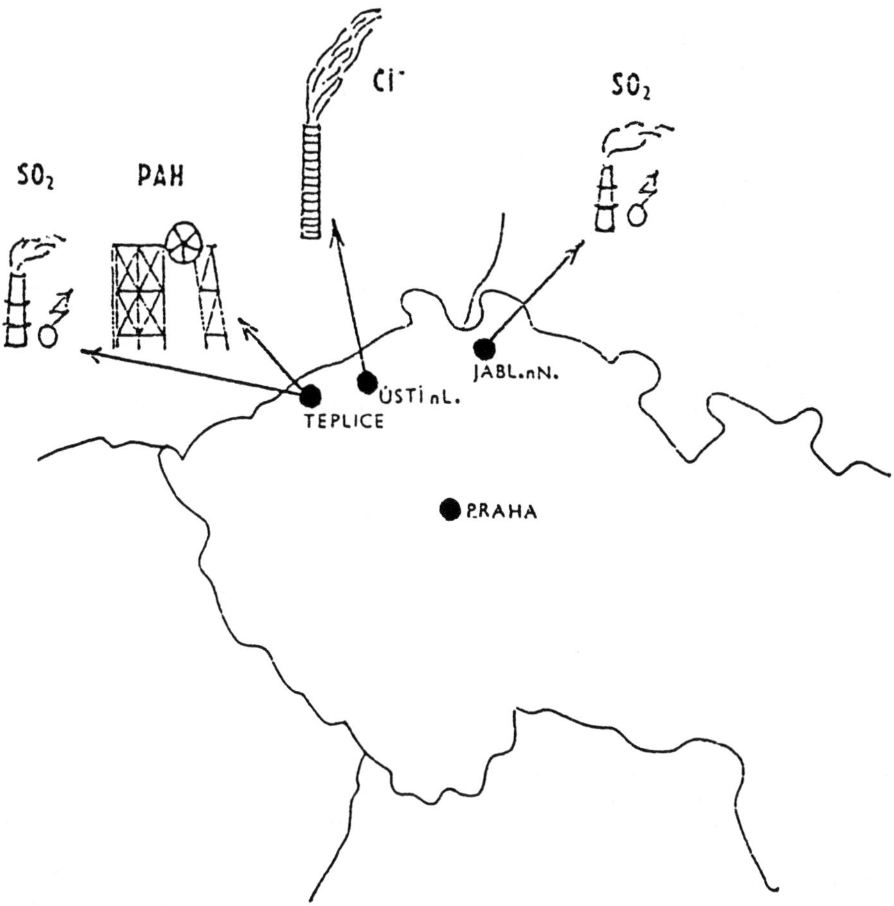

Fig. 1. Air pollution sources in Northern Bohemia. SO_2-brown coal power stations, PAH-brown coal strip mines, Cl^--chlorine chemical industry.

Information about pregnancy outcomes was obtained from medical records at maternity hospitals. Diagnosis of CGA according to the International Classification of Diseases (ICD; WHO 1977) was verified at 1 year of age. The study lasted for the period 1982–1986; altogether 18,060 pregnancy outcomes were examined. As markers of environmental pollution, data on sulphur dioxide and nitrogen oxides were obtained from monitoring stations (Šrám et al. 1990a,b).

We tabulated the total number of pregnancies and the percentage of abnormalities: district A, 7644 pregnancies, 9.8 % CGA; district B, 7190 pregnancies, 8.2 % CGA; district C, 3226 pregnancies, 6.4 % CGA. When we analysed the frequency of CGA, there were, significantly higher levels in both mining districts (Table 1). Similarly, in these same districts there more children were born with weights lower than 2500 g at delivery (Table 2). When the spectrum of CGA in each district was compared with data from the USA and Hungary, districts A and B showed a tendency toward a higher frequency of CGA of the cardiovascular system, urinary system and genital organs and of integument (Table 3).

Table 1. Frequency of congenital anomalies[a]

District	Year				
	1982	1983	1984	1985	1986
A	11.1	9.7	9.9	9.1	8.7
B	8.5	7.9	7.8	7.8	8.7
C	6.7	6.0	6.5	–	–

[a]% of total births

Table 2. The frequency of children with the weight lower than 2500 g at delivery[a]

District	Year				
	1982	1983	1984	1985	1986
A	8.1	8.4	7.7	7.5	8.7
B	8.3	8.3	9.2	7.9	6.5
C	5.5	6.5	4.3	–	–

[a]% of total births

Table 3. The spectrum of congenital anomalies in Northern Bohemia, USA (Mysianthopoulos and Chung 1974) and Hungary (Czeizel and Sankaranarayanan 1984)[a]

Anomaly of:	District A	B	C	United States	Hungary
Brain and back bone	0.4	0.4	0.3	0.5	0.3
Eye	0.3	0.3	0.2	0.24	0.03
Ear, face and neck	0.5	0.5	0.5	0.4	0.2
Cardiovascular	1.2	1.0	0.6	0.86	0.8
Respiratory	0.1	0.1	0.2	0.14	0.03
Digestive	0.2	0.3	0.2	0.62	0.28
Urinary and genital	1.8	1.9	1.1	1.15	0.94
Skeletal	4.0	2.3	3.1	4.38	3.14
Integument	1.9	1.8	0.7	1.0	0.08
Others	0.8	0.5	0.5	0.13	0.21

[a]% of total births

The effect of SO_2 exposure on CGA, related to the month of conception, was compared in district A in 1982 (Table 4). Based on these results it was assumed that SO_2 increased the frequency of CGA in the population. But when the same data were analysed for the year 1984, no such relationship was observed.

Sulphur dioxide was used as a marker of air pollution. Its concentration was compared to the number of newborn children in each month and to the frequency of CGA in district B. Spectral analysis confirmed that there is a strong circa-annual periodicity in the SO_2 data.

Table 4. Effect of SO_2 to CGA, related to the month of conception in district A in 1982

Month	SO_2 ($\mu g/m^3$)	CGA (%)
1	492	13.7
2	344	12.3
3	168	14.6
4	114	12.2
5	78	13.1
6	48	9.2
7	56	9.9
8	69	5.7
9	99	5.8
10	136	8.0
11	165	11.4
12	224	13.6

We tried to detect whether the SO_2 annual component induces a driving effect in the month's frequency of CGA. For this purpose we used the coherence function, which indicated some, but not much, convincing connection between SO_2 and CGA. We also tried to explain the dynamics of CGA in terms of models with intervention variables consisting of the instant values of SO_2 and their time-lagged twins. In these models the CGA rate was adjusted to the date of conception. None of these models explained more than 10 % of variance of CGA. This is probably due to the fact that CGA do not exhibit any circa-annual periodicity.

Later we turned our attention to the analysis of conceptions of children born in district B. Birth frequency in Bohemia has a great annual periodicity which explains more than 63 % of its variance. The corresponding data from district B do not have such a strong annual periodicity; they explain only 9 % of the variance of the month's birth dynamic.

This prompted us to focus attention on the phase relationship between SO_2 and the frequency of conceptions in district B. Whenever SO_2 is above 200 $\mu g/m^3$, the conception frequency wave-curve is declining (see Fig. 2). Simultaneously, the annual components of birth frequency in district B attain their maximum levels a month before the maximum in Bohemia. This finding could contribute to the explanation of why the annual periodicity in district B is more than seven times smaller then the amplitude of the annual periodicity in the whole of Bohemia.

We can speculate that the increase of SO_2 suppressed the number of successfully fertilized oocytes. We hypothesize that other mutagens inducing mutations in gametes were potentiated by SO_2 due to the suppression of the DNA repair mechanism. Other reasons may be the induction of chromosome aberrations by SO_2 and other pollutants. Oocytes carrying such changes are not implanted or they are aborted during very early stages of pregnancy. We tried to relate these results to the number of days when the level of SO_2 was higher than the maximum allowable concentration (MAC), which amounts to 150 $\mu g/m^3$ over 24 hours in Czechoslovakia. If the level of SO_2 was two times higher than MAC

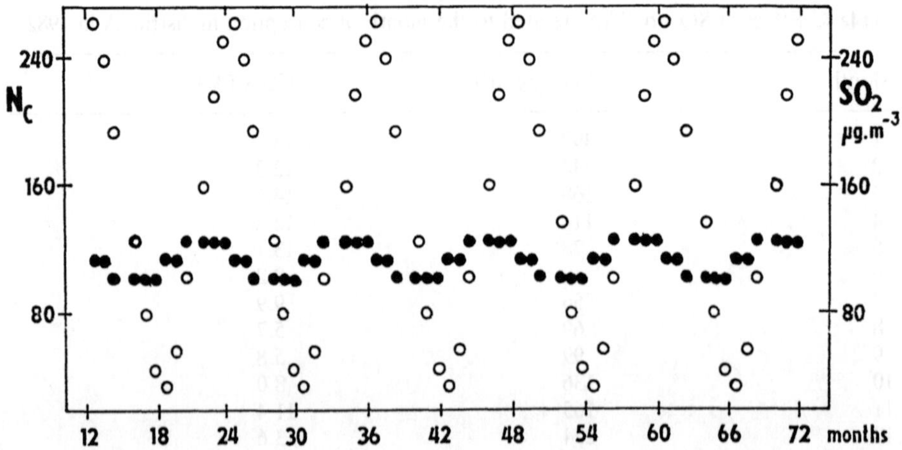

Fig. 2. Annual periodicity of conceptions (●) and SO_2 (o) at Teplice. N_c, number of conceptions per month.

during 10 days in a month, a tendency toward an increase of CGA was observed.

It should be stressed that observed changes were not only the results of SO_2 exposure. They were combined with NO_x, but the annual periodic component of NO_x was not as intensive as that of SO_2. The brown coal from Northern Bohemia also contains heavy metals, which may also suppress DNA repair.

These results may be in agreement with the ideas of Jelínek (1988; Fára et al. 1988), who studied problems of dose and injury in teratology. Developing embryos may react to a harmful action by dying, by structural and/or functional anomaly, or by growth retardation. All three events are understood as results of embryotoxicity.

What is the relationship between the dose and effect? If the dose increases, only the number of dead embryos increases. The number of malformed fetuses increases during the first step dose increases but later, paradoxically, decreases. Experiments have shown that malformed fetuses died because the injuries induced by the higher doses were not compatible with life. Therefore, malformed fetuses get placed in the category of dead fetuses. This is the reason Jelínek does not accept the category of CGA as the best endpoint for the evaluation of environmental pollution.

Other important points are the critical and sensitive periods in the human development. The most important period is between the third and eighth week of intrauterine life. This period is critical for the development of external shape and formation of various organs and systems. If embryonal development is affected in this period, gross malformations may be induced, such as anencephaly, spina bifida, cleft palate, or CGA of the limbs and heart.

From the ninth weeks to delivery is the period of fetal development when each organ grows and differentiates. Changes induced in this period especially affect the functioning of these organs. Environmental factors can affect brain maturation in the period of delivery; these effects may, in the course of the development of such an individual, result in abnormal behaviour. This finding is crucial for the problem of how environmental pollution, especially by mutagens, can affect human populations (Jelínek 1988).

When we analysed our data from polluted region in Northern Bohemia, we drew the following conclusions:

1. The frequency of CGA is approximately 4–5 times higher in this region than in other parts of Bohemia, according to official records.
2. The increase of CGA of the cardiovascular, urinary and genital systems and integument corresponds to Jelínek's idea about critical and sensitive periods.
3. Approximately 60 % more children were born with weights lower than 2500 g at delivery in the mining districts than in other districts of Bohemia. This is reflected in more children with immature brains, who may later suffer from functional changes that will probably affect their behaviour.
4. Analyses of annual periodicity of conceptions and SO_2 indicate that high exposure to pollutants may affect the first period of embryonal development and thereby decrease the number of conceptions ending in the delivery of child.

Combining our results with other data from the literature, we propose a hypothesis corresponding to Jelínek's experimental results: Where there is a region with an increased frequency of CGA, we may also expect an increase in functional changes in the population of the region. To analyse brain functional changes, it is convenient to analyse minimal brain dysfunctions (MBD; Gebhart et al. 1990).

MBD diagnosis in children is related to mental retardation, developmental disorders, and hyperkinetic and behavioral dysfunctions (Tyl et al. 1990). There may also be significant impact of MBD in the adult. It may speed up the process of aging and increase the frequency of psychosomatic diseases and the sensitivity to alcoholism.

Are there any experimental data which would indicate that this idea is plaussible?

Environmental Factors Affecting Behavioral Function

Behavioral function may be affected in two ways: 1) through exposure to pollutants during neonatal development or 2) by affecting gametes and transmitting induced changes to the progeny. These data have been obtained only over the last several years, but they nevertheless indicate that behavioral changes induced by pollutants may be a new, still not very well recognized problem (Table 5).

Neonatal Development

The effect of environmental contaminants on behavioral alterations was studied by Singh (1986, 1988, 1989). Pregnant mice were exposed during the 7th – 18th

Table 5. Environmental factors affecting behavioral functions (in mammals)

Sulphur dioxide	Singh (1989)
Nitrogen oxide	Singh (1988)
Nitrous oxide	Friedler and Meadows (1987)
Carbon monooxide	Singh (1986)
Mutagens – cyclophosphamide	Hsu et al. (1987)
	Auroux et al. (1990)
Lead	Brady et al. (1975)
Industrial solvents (2-methoxyethanol)	Nelson et al. (1984)
Ethylenoxide	Barnett et al. (1988)
Ionizing radiation	Schroeder (1978, 1979, 1985); Lowery et al. (1990)

day of pregnancy to 32 or 65 ppm of SO_2 (Singh 1989), to 22 or 45 ppm of NO_2 (Singh 1988), or to 65 or 125 ppm of CO (Singh 1986). Exposure to SO_2 decreased mean pup weight and increased time the required for the righting reflex and negative geotaxis. Exposure to NO_2 decreased birth weight and increased the time required for the righting reflex and aerial righting score. If pregnant mice were exposed to CO, their pups required increased times for the righting reflex and negative geotaxis as well as aerial righting score. All these results indicate significant functional deficits in the progeny of mice.

Mutagens

Determining mutagens may affect behaviour is quite a new field. It can be studied as male-mediated effects. It is expected that mutations may lead to the expression of functional abnormality, such as abnormal behaviour. Behavioral effects are very sensitive endpoints. They can be indicators of a toxic action; subtle disturbances in behaviour may give clues to the mechanisms by which the CNS is affected. The behavioral abnormalities may occur in the absence of gross morphological defects in the offspring, or at exposure doses well below those used to elicit positive responses in classical mutagenicity assays. The mechanisms or genes involved in such behavioral aberrations remain unknown. However, if these effects are transmitted to the F_2 generation, a genetic basis for mutation is strongly suggested (Adams et al. 1984).

Pioneering results about the effect of mutagens on the CNS were obtained by Lohman et al. (1990) using the new methodology of recombinant DNA with a marker gene, Lac Z, located on the bacteriophage lambda shuttle vector and transgenic mice. They observed a dose-dependent increase of Lac Z mutations isolated from brain cells of transgenic mice one week after treatment with ethyl nitrosourea (ENU). Simultaneously, they observed that these mutations were organ-specific and that the brain was the most sensitive organ.

The transmissible effect was proved in rats who were treated with a daily dose of 10 mg of cyclophosphamide (CY)/kg for 15 days and mated 100 days after treatment with virgin females. In the F_1 generation treated by CY, decreased learning capacity (avoidance response in shuttle box) and decreased spontaneous activity (open field) were observed in male offspring. Hippocampal choline acetyltransferase (CAT) activity was also significantly decreased. In the F_2 generation of a subgroup of F_1 parents who were unsuccessful in the learning test, there was an increase in postnatal mortality and reduced spontaneous activity. In subgroups of F_1 parents that were either successful or unsuccessful in the learning test, the learning ability of F_2 males was significantly impaired. Biochemical assays showed decreases of CATactivity and norephinephrine from the frontoparietal cortex. The learning deficit was still observed in the F_3 generation (Auroux et al. 1990).

Brady et al. (1975) reported that paternal lead exposure affected the learning ability of the offspring. Inhalation exposure to 3-methoxy-ethanol in pregnant rats induced differences in avoidance conditioning. Paternal exposure led to neurochemical alterations in offspring (acetylcholine, dopamine, norepinephrine; Nelson et al. 1984). Following acute NO exposure to paternal rats, the offspring demonstrated delays in developmental indices and neuroendocrine responses (Friedler and Meadows 1987). Ethylenoxide exposure to male mice induced neurological defects in their progeny. Carriers of behavioral mutations also have a specific chromosome translocation ([t(17; 4)]; Barnett et al. 1988).

Ionizing radiation is associated with known behavioral defects. Schroeder (1978, 1979, 1985) demonstrated alterations in learning ability and male aggressiveness in the offspring of irradiated male mice. Lowery et al. (1990) studied the effect of 0–300 rad acute whole-body irradiation. Results from offspring revealed significant dose response patterns in cliff avoidance, active avoidance and swimming development.

This short review indicates that data from recent experiments fully agree with the idea that behavioral changes can be induced by environmental mutagens. The nervous system may show great susceptibility to the effect of mutations, which may be related to the great number of genes which are expressed in nervous tissue.

Population Studies

Studies about the effect of mutagens on human populations and mental health are extremely scarce.

Otake and Schull (1984) demonstrated that the fetal brain is highly sensitive to atomic bomb irradiation. Their Hiroshima-Nagasaki study examined 1600 surviving children exposed in utero. They observed a reduction in IQ by 0.2–0.4 per rem if the brain was affected during the most sensitive period of 8–15 weeks post fertilization.

In view of the geographical and seasonal variations in radioactive fallout in Norway during and following the atomic bomb test period 1954–62, it was hypothesized that a radiation effect might be discernible by comparing temporal patterns of intellectual characteristics in corresponding cohorts from different regions of Norway. During that period Norway received relatively high levels of radioactive fallout from nuclear bomb tests. Judging from the content of ^{137}Cs in milk, there were strong seasonal and regional differences, with the West-facing coast receiving 5–10 times as much radioactivity as the inland and Eastern parts.

Oftedal (1988) studied the scholastic achievements in the cohort of children born in 1965, according to centrally organized norm-related tests for pupils in the 7th, 8th, and 9th grades of public school. He analysed 13,000 individual test results in the subjects of Mathematics, Norwegian and English. Oftedal observed a clear deficit in the "West" cohort born in the early months of the year, the originally hypothesized results of a fallout radiation insult. This deficit could be attributable to insufficient repair gene activation. At least five repair systems are known, and in the mammalian genome as many as 100–200 genes may be involved. There is evidence from several organisms that a significant portion of the repair capacity is held in abeyance and is turned on only when the need arises. A close adjustment between the level of repair system activity and the prevailing level of background noxes could be predicted. Oftedal (1988) interpreted his results in such a way, proposing that the increase in radiation was not capable of inducing the requisite effective repair capacity. This might be due either to the time lag before repair functions could become available or because the level of insult was not sufficiently changed. But if the constitutive repair mechanism cannot cope immediately with the new level of insult, a period of greatly enhanced sensitivity might follow.

It is extremely difficult to analyse the effects of environmental pollution on mental health, because the same effect can be induced by various factors. Nevertheless, a comparison of the morbidity of children (7–15 years old) in the industrially polluted region of Northern Bohemia with the morbidity in the Czech Republic indicates that there are approximately twice as many mentally ill and retarded (Table 6) children in these polluted regions.

Table 6. Morbidity of children (7–15 years) in the industrially polluted region of Northern Bohemia (Czechoslovakia)

Disease	Number of cases/100 children	
	Czech Republic	Polluted Region
Urinary and kidney	1.42	1.68
Respiratory	0,45	1.40
Mental	2.00	4.09
Endocrine	1.17	1.54
Dermatic	0.73	1.09
Others (chronic)	0.92	1.79

Impact of Environmental Pollution on Mental Health

Our review indicates that environmental mutagens and other pollution factors may significantly affect brain development. This may occur during the critical and sensitive periods of CNS development as well as during the third trimester, when induced changes may correspond to various behavioral defects – patterns varying from learning ability and IQ to various other forms of minimal brain dysfunction as well as expressed mental illness.

This new knowledge should be used for preventive measures when etiological factors are known. More serious are the indications that brain injury is induced by mutagens in gametes. Based on the results obtained in mammals, we can assume that mutations in gametes can accumulate from generation to generation. Behavioral changes induced in the parental generation may be potentiated, if people with the same defects from the same region have children. This means that, in the long term, this type of injury may potentiate in time, resulting in decreasing IQ and various behavioral dysfunctions.

Simultaneously, the mental health of parents may be improved by receiving better information about the impacts of environmental pollution on the health of their children. In some cases their mental health may be affected by the continual stress they feel due to their inability to improve their children's environment.

This situation raises an extremely important ethical question. How should we solve the above-mentioned problems? It is clear that living in a highly polluted region can affect the mental health of a population for at least the next two generations. Even if this pollution is decreased, the health effects of induced injury will last for more than the next 100 years (two generations + life expectancy). This is a great burden to the health of a population, probably even more significant than anybody can seriously evaluate.

Usually in such regions there is a shortage of resources to improve the environment in a relatively short time. Therefore, the population will be exposed to heavy pollution for several more years and all injuries will increase. However, now that we know about this impact of environmental pollution on human health, we should try to stimulate action and promote some preventive measures.

WHO is now discussing the new field of ecopsychiatry. Under this new discipline, epidemiological studies should be started in highly polluted regions (with chemical mutagens or ionizing radiation) and analyze how these pollutants may affects, for example, minimal brain dysfunctions and school achievements in children. Simultaneously, new achievements in the field of antimutagenesis can be used to try to suppress these harmful effects using antimutagens. Probably the highest priority should be given to pregnant mothers in such regions.

I believe that it is ethically imperative to start such studies with the aim of working out effective preventive measures.

References

Adams PM, Shabrawy O, Legator MS (1984) Male-transmitted developmental and neurobehavioral deficits. Teratogenesis Carcinog Mutagen 4:149–169

Auroux M, Dulioust E, Selva J, Rince P (1990) Cyclophosphamide in the Fo male rat: physical and behavioral changes in three successive adult generations. Mutat Res 229:189–200

Barnett LB, Lewis SE, Davisson M (1988) Genetic studies of a behavioral mutant induced by ethylene oxide in the mouse. Environ Mutagen (Suppl 11) 11:9

Bochkov NP, Šrám RJ (1989) Modern trends and realities of genetic monitoring. Biol Zentralbl 108:341–348

Brady K, Herrera Y, Zenick H (1975) Influence of parental lead exposure on subsequent learning ability of offspring. Pharmacol Biochem Behav 3:561–565

Czeizel A, Sankaranarayaman K (1984) The load of genetic and partially genetic disorders in man. I. Congenital anomalies: estimates of detritment in terms of years of life lost and years of impaired life. Mutat Res 128:73–103

Fára M, Jelínek R, Peterka M, Dostál M, Hrivnáková J (1988) Orofacial clefts. A theoretical basis for their prevention and treatment . Charles University, Prague,

Friedler G, Meadows ME (1987) Paternal exposure to nitrous oxide: effects on development of mouse offspring. Neurotoxicity 7:365

Gebhart JA, Dytrych Z, Tyl J, Šrám RJ (1990) On the incidence of minimal brain dysfunction syndrome in children (in Czech). Čs Psychiat 86:1–6

Hsu LL, Adams PM, Fanini D, Legator MS (1985) Ethylene dibromide: effects of paternal exposure on the neurotransmitter enzymes in the developing brain of F_1 progeny. Mutat Res 147:197–203

Hsu LL, Adams PM, Legator MS (1987) Cyclophosphamide: effects of paternal exposure on the brain chemistry of the F_1 progeny. J Toxicol Environ Health 21:471–481

Jelínek R (1988) The principles of teratogenesis revisited. Congenital Anom (Suppl) 28:145–155

Lohman PHM, Vogel EW, Morolli B, van Zeeland AA, Vrieling H (1990) Development and application of new methodologies applicable to research on complex environmental mixtures. (in press)

Lowery MC, Au WW, Adams PM, Whorton EB, Legator MS (1990) Male-mediated behavioral abnormalities. Mutat Res 229:213–229

Mysianthopoulos NG, Chung CS (1974) Congenital malformations in singletons: epidemiological survey. Birth Defects 10:1–58

Nelson BK, Brightwell WS, Burg JR, Massari WJ (1984) Behavioral and neurochemical alterations in the offspring of rats after maternal or paternal inhalation exposure to the industrial solvent 2-methoxyethanol. Pharmacol Biochem Behav 20:269–279

Oftedal P (1988) Scholastic achievement in relation to fetal exposure to radioactive fallout in Norway. 14th LH Gray Conference, Oxford, pp 345–353

Otake M, Schull WJ (1984) In utero exposure to A-bomb radiation and mental retardation: a reassessment. Br J Radiol 57:409–414

Schroeder JH (1978) Water-escape performance and water escape learning ability in laboratory mice as dependent of different gonosomal constitution and paternal irradiation of spermatozoa with 600 R of gamma rays. Ber Naturwiss Med Ver Innsbruck 65:163–188

Schroeder JH (1979) Hereditary changes of male aggressiveness after gamma irradiation of mouse spermatozoa with 600 R of gamma-rays as dependent on fertility. Ber Naturwiss Med Ver Innsbruck 66:131–137

Schroeder JH (1985) Increase in aggressiveness of male mice after irradiation of mouse spermatozoa with 600 R of gamma-rays as dependent on fertility. Behav Genet 10:387–400

Singh J (1986) Early behavioral alterations in mice following prenatal carbon monoxide exposure. Neurotoxicology 7:475–482

Singh J (1988) Nitrogen dioxide exposure alters neonatal development. Neurotoxicology 9:545–550

Singh (1989) Neonatal development altered by maternal sulphur dioxide exposure. Neurotoxicology 10:523–527

Šrám RJ, Binková B, Gebhardt JA (1990a) Impact of environmental mutagens on mental health. In: Bulyzhenkov V, Christen Y, Prilipko L (eds) Genetic approaches in the prevention of mental disorders. Springer, Berlin Heidelberg New York, pp 74–80

Šrám RJ, Rožníčková I, Albrecht V, Beránková A, Machovská E (1990b) Monitoring congenital anomalies in populations exposed to environmental mutagens. In: Kappas A (ed) Mechanisms of environmental mutagenesis - carcinogenesis. Plenum, New York, pp 255–266

Tyl J, Matějček Z, Albrecht V, Beránková A, Kotěšovec F, Machovská E, Reifová L, Dytrych Z, Gebhart JA, Šrám RJ (1990) Epidemiological investigation of signs of minimal brain dysfunctions. I. Design and testing of method-incidence of signs in the investigated groups (in Czech). Čs Psychiatr 86:27–36

World Health Organization (1977) International classification of diseases. Manual of the international statistical classification of diseases, injuries and causes of death, vols 1 and 2. WHO, Geneva

A Molecular Genetic Approach to Neuropsychiatric Disease – Some Technical and Ethical Considerations

Th. Friedmann and *F.H. Gage*

Most human traits, both normal and pathological, are determined by interactions between genetic and non-genetic environmental factors. Some of the genetic factors are simple, single gene factors whereas many others involve several or possibly many genes interacting with each other as well as with the environment. It is also justified to assume that all normal and pathological functions and characteristics of the human central nervous system, ranging from relatively simple phenotypes to the most complicated of human behaviors, are certain to be as much a combination of genetic and environmental factors as are all other aspects of human biology and disorders of all other organs. It seems evident that a degree of balance between the biological and environmental influences will be found not only to determine behavioral aspects of genetically simple neuropsychiatric disorders such as Lesch Nyhan syndrome and Huntington's disease – diseases in which genetic factors almost certainly predominate – but also to contribute in major ways to more complex disorders such as schizophrenia and other affective illnesses in which genetic components have not yet been clearly characterized. Furthermore, it seems inevitable that complicated and probably multiple genetic factors interacting with environmental factors will be associated with degenerative defects such as those responsible for Parkinson's and Alzheimer's diseases and other human neurological disorders. Ultimately, aspects of even the most inaccessible and possibly the most unique of human traits – intelligence, memory, cognition, even creativity – must have genetic as well as environmental components. Through a combinantion of traditional biochemical genetic techniques and molecular genetics approaches, those genetic components will inevitably begin to be identified and eventually understood.

Traditional "Forward" Genetics

Some of the most impressive accomplishments of modern molecular genetics include the development of tools and techniques by which genetic components of human traits can be identified, characterized and understood. The traditional approach to such an understanding requires the demonstration of the biochemical and metabolic components of a genetic characteristic, whether it be a normal metabolic step or a disease-causing defect. Once the biochemical properties and

Srám, Bulyzhenkov, Prilipko, Christen (Eds.)
Ethical Issues of Molecular Genetics in Psychiatry

functions of a gene product are known, it is possible in principle to use one of the many cloning strategies to isolate the relevant gene and study its structure and the relationship of its structure with function. This traditional approach has come to be called "forward" genetics, and it has led to the isolation of genes that regulate important metabolic pathways and that encode vital genetic products and functions involved in important genetic disorders, many of them representing the classical "inborn errors of metabolism." This is a valuable approach to the understanding of many human diseases, but it suffers from a number of severe limitations. The most serious is the fact that in this scheme one must know the biochemical or metabolic function encoded by a gene before one can devise methods to isolate and characterize the gene itself. This respresents a major drawback to an understanding of most human traits and disorders. Because there has been so little rigorous understanding of the causative biochemical defects underlying most neuropsychiatric disorders, this traditional "forward" approach has not yet been very effective for the elucidation of their genetic bases or for the development of definitive treatments.

"Reverse" Genetics

The development of molecular-based linkage analysis during the past decade has given birth to the concept of an alternative approach, "reverse genetics." Unlike the limitations of more traditional "forward" genetics, this approach allows, in principle, the isolation and genetic understanding of phenotypes and disease traits even in the total absence of any demonstration of the biochemical or metabolic nature of the defect. The "reverse" genetic approach is based on the use of genetic "walking" or "jumping" techniques to move from a sequence near a disease locus on a chromosome to the locus itself. These steps require absolutely no knowledge of the biochemical and metabolic consequences of the genetic defect. The recent isolation of the gene for cystic fibrosis (Riordan et al. 1989) is the first instance of the isolation and characterization of a human disease-related gene entirely by this kind of approach. Still newer approaches for characterizing a number of simultaneously acting quantitative traits have now also come to be applied to complex traits in lower organisms, and while it certainly will be some time before they are applied extensively to human traits, it seems likely that some complicated multigenic human traits – both normal developmental characteristics such as intelligence and elements of personality as well as disease phenotypes such as hypertension, cardiovascular disease, some forms of cancer and others – will eventually be approached and understood at the genetic level.

Molecular Genetics and Nueropsychiatric Disease – Genetic Mapping

Without doubt, the genetic revolution will lead to an understanding of some of the mechanisms by which genetic defects cause, or contribute to, psychiatric disorders and psychotic and other aberrant behaviors. These genetic factors will, slowly but inexorably, come to be identified and understood, many through the tools and techniques of reverse genetics. As nearby linked markers are identified, genes related to CNS diseases will come to be isolated, slowly at first but then at an increasing pace, just as has occurred with the gene for cystic fibrosis. Early promising results have already been obtained through the isolation of DNA sequences genetically linked, and therefore physically close, to genes responsible for previously inaccessible neuropsychiatric disorders such as Huntigton's disease (Gusella et al. 1983), schizophrenia (Sherrington et al. 1988) and Alzheimer's disease (St. George-Hyslop et al 1987). The first instance of a genetic linkage to a neuropsychiatric disease came in 1983 with the demonstration of the linkage of a DNA probe isolated from near the telomere of human chromosome number 4 to the gene responsible for the dominant Huntigton's disease. However, despite many hundreds of person years since then, the gene itself has not yet been isolated, giving a good indication of exactly how difficult this kind of genetic detective work still is. Linkage of sequences from chromosome 21 to a genetic factor associated with Alzheimer's disease and of other sequences from chromosome 5 to schizophrenia in some pedigrees lends encouragement to the concept that genes associated with these and other severe and burdensome disorders will eventually be identified and understood, despite the likelihood that they will be found to be multigenic and heavily influenced by environmental factors as well.

Crucial to the development of these concepts and to the application of techniques of reverse genetics is the development of thorough genetic and physical maps of the human genome (Friedmann 1990), projects that are well underway now in many countries under the banner of the Human Genome Organization (HUGO) and others of which are organized through major national human genome programs in the United States, Japan, Great Britain, the Soviet Union, Italy and other countries. The combination of continued traditional forward genetic approaches with the explosion of new knowledge derived from mapping and reverse genetic programs will produce immense amounts of new genetic information about many normal and pathological human conditions and characteristics. Obviously, it is inevitable that much of this new knowledge will deal with the functions and malfunctions of the human central nervous system. For the very first time, it seems possible to begin to understand the mechanisms of many normal and abnormal functions of the human brain, and to enter a new era of far more efficient methods of treating the neuropsychiatric disorders that have generally been intractable and unapproachable.

While the amount and quality of this new information will represent a quantum leap in our understanding of human biology in general and of the human brain in particular, it also poses important ethical and public policy problems that we must anticipate and solve before embarking on broad new programs aimed at the identification, characterization and treatment of neuropsychiatric disease. We must know what to do with so much new knowlege, and how to ensure that our profound new genetic understanding of human diseases is used efficiently for effective disease prevention and treatment. We will be faced more acutely than ever with the problems of making enlightened, humane and acceptable distinctions between genetically determined aberrations in cognition and behavior and the other environmental causes of deviant behavior, of distinguishing between therapy of disease and the non-therapeutic manipulation or enhancement of other characteristics.

Diagnosis – Uses of Molecular Genetic Information

The most immediate impact of molecular genetic knowledge on neuropsychiatric disease will almost certainly be to develop a rigorous and molecularly based classification and etiological scheme for the disorders, to develop effective means of identifying factors that cause or predispose to neuropsychiatric disease, to identify patients with genetic neuropsychiatric diseases or those at risk, and to develop more effective genetic counselling procedures and prevention strategies. Diagnostic and screening information will obviously be used first in families in which disease has already occurred, in targeted populations particularly susceptible to the disease and, eventually, in entire populations. With a precise identification of the aberrant biochemical and genetic step, the more rational design and development of pharmacological agents will become more effective and specific.

Therapy – Improved Approaches to Traditional Pharmacological Therapy

With an increasing understanding of the genetic components and mechanisms of neuropsychiatric disease will come an increasing opportunity and obligation to treat disease; in the case of neuropsychiatric disease, this will mean the use of drugs to alter behavior. In most kinds of genetic disease, there is and will continue to be a great discrepancy between the exploding capability to detect and predict disease and our ability to devise effective treatments. Oddly enough, the situation seems somewhat different in the case of neuropsychiatric disease. It is fair to say that the use of some kinds of psychoactive drugs and tranquilizers has revolutionized the management of many kinds of psychiatric diseases, even though much of today's sophisticated understanding of neurochemistry seems not yet to have been translated into a precise understanding of mechanisms of neuropsychiatric pathology. The impressive advances that have taken place in our understanding of the neurochemical basis of brain function and in the design of psychoactive drugs during the past several decades make it obvious that rational design and development of pharmacological interventions based on knowledge of underlying neurochemical defects wil improve markedly. But what may seem to be rational therapy based on knowledge of neurochemical aberrations may really represent a blunt pharmacological assault on only one aspect of a defective biochemical pathway in an affected brain. In most cases, we cannot yet aim therapy at the precise step known to be responsible for the dysfunction. Therefore there are likely to be many instances in which pharmacological approaches to therapy will be, at best, only partly effective and overly blunt. In some cases there will be a need for more precise, targeted, rational methods to correct the underlying defect, the aberrant genes.

Gene Therapy

During the past several decades, it has become clear that, until the development of the concepts and tools of molecular genetics, Medicine never had the luxury of developing therapies aimed directly at underlying genetic defects. Rather, we had to be content with treatments directed instead at the metabolic consequences of genetic defects. That approach is a legacy of two giants of genetics whose ghosts still pervade all discussion of human genetics: Gregor Mendel, whose presence is so tangible here in Brno, and the English physician Sir Archibald Garrod. Since the discovery of the rules of genetic inheritance by Mendel in the middle of the 19th century and the first application of those rules to an understanding of the biochemistry and the rational treatment of human genetic disease by Garrod in England at the beginning of the 20th century (Childs 1970), Medicine has depended on therapies based very largely on manipulation of metabolic pathways: replacement of vital metabolites, removal of toxic metabolites, dietary restriction to prevent accumulation of toxins, restoration of enzymic activities by providing co-factors to defective apoenzymes or even possible replacing functional enzymes. In no cases have we had the luxury of fixing what really was broken, the gene.

The recent revolution in molecular and human genetics has now provided tools and concepts that promise to overturn all of the more classical current approaches to therapy. We now have the opportunity to imagine treatments aimed directly at the restoration or correction of the defective genes themselves: gene therapy (Friedmann 1989). During the past decade a number of very informative model systems have been developed, and the rate of progress in this field has been extensively documented. Several very important conclusions can be derived from the recent history of progress with these model studies. The first is the realization that techniques for restoring genetic functions that may have seemed so remote and even so inconceivable technically until recently are now quickly falling into our grasp. We are now able to introduce foreign genes stably into defective cells, to complement genetic defects and correct disease phenotypes in vitro and soon in vivo. Furthermore, a very high degree of acceptance has now been achieved in the scientific, public policy, ehtics and medical communities toward the goals and techniques of gene therapy. In fact, the new molecular genetic concepts of the specific correction of genetic diseases rather than mere amelioration of symptoms are now coming to be so pervasive as to represent a true and astonishing revolution in biomedical thinking and in the design of therapy for so much human disease.

Genetic Approaches to Treatment of Central Nervous System Disease

Definitive genetic correction of aberrant central nervous system (CNS) phenotypes might theroretically be achieved through the introduction of foreign normal genes directly into defective neural cells with viral vectors or other gene transfer tools. While this approach is ideal in principal, it is not yet feasible for a number of reasons. The currently available kinds of efficient viral gene transfer vectors, the retroviral vectors, are incapable of infecting non-replicating, quiescent or post-mitotic cells, and therefore, under most circumstances, they cannot be used for the genetic modification of genetic defects in neurons. Other kinds of vectors have not yet been developed for the stable expression of foreign genes in neurons, although early results with prototype herpesvirus-based vectors are promising and may eventually lead to useful tools to introduce new genes into neural cells. Non-viral gene transfer techniques, such as CNS-targeted liposomes, may also eventually be useful, although it will probably be important to understand and be able to modify the transport of viruses and other macromolecules across the blood-brain barrier before many such gene transfer techniques are applicable to the mammalian CNS.

But there are alternative approaches to a genetic correction of genetic defects of the mammalian CNS, and we now know that at least some functions of the mammalian CNS can be modified by these other genetic means. Using the combined tools of neurosciences and molecular genetics, our colleagues and we have proposed and studied the modification of several CNS functions and neurological phenotypes by a combination of retroviral-mediated gene transfer and cell grafting to the brain (Gage et al. 1987). We have introduced normal foreign genes into non-neural cells in vitro and grafted the resulting genetically modified cells into recipient rat brains at a location in which the production and diffusion of a gene product to nearby defective resident neural cells is sufficient to complement a cellular or genetic defect. We have concentrated on two principal models, those involving the ameliorative or protective role of nerve growth factor (NGF) in the degeneration of cholinergic neurons as a possible therapeutic approach to Alzheimer's disease (Rosenberg et al. 1988), and the complementation of the dopamine neurotransmitter defect associated with Parkinson's disease in humans (Wolff et al. 1989). These early studies have established the feasibility of this approach to neuropsychiatric disease in which foreign genes expressed by grafts of genetically modified cells in the mammalian brain can complement defects in cellular functions in several animal models of human CNS disease – a form of indirect gene therapy.

Ethical Considerations

Without doubt, advances in molecular genetics will lead to powerful methods for the modification of human behavior. It is very important to determine under what circumstances we wish to undertake such behavior modification steps and what their ethical costs may be.

Most, but certainly not all, human societies embrace the goals of Medicine, whether it be primitive or highly technological medicine, to alleviate human suffering caused by disease. We accept the principle of active intervention in the natural course of human health and look upon injury and illness as evils to be conquered. This attitude tacitly accepts the concept of errors in nature – mistakes in the flow of the genetic information from one human generation to another. It rejects the acceptance of the results of a genetic and a biological lottery that presents human beings with an unalterable fate. We wish and strive for better for those around us who suffer from disease. This argument is not meant to suggest that other attitudes to our universe are any the less acceptable in other cultures. A number of Oriental views of the universe and the interaction of human beings with the universe teach that it is the rejection of the world as it is, the yearning for things better, that is at the root of human unhappiness. Contentment comes with an ability to flow with the forces of nature and accept its whims – its detriments and injuries as well as its bounties. That, of course, is in conflict with the goals of Medicine.

In Western cultures, most moral agents have expressed little or no disagreement with the need for specific kinds of genetic modifications, pharmacological or genetic, of some specific aberrant human behavioral characteristic. Most in our society would agree readily that the psychiatric aspects of genetic diseases such as Huntington's disease should be prevented or corrected; they are the results of disease, clearly abnormal and not considered by our societies to be in the best interest of the patients. There is almost universal agreement that most pharmacological approaches to the restoration of normal function in such instances are consistent with the principles of Medicine. Similarly, the aberrant behaviors of many patients with schizophrenia, the personality destruction in patients with Alzheimer's disease, the incapacitation of patients suffering from Parkinson's disease, the retardation in patients with Down, fragile X and Lesch-Nyhan syndromes and other similar diseases would be looked upon by most of us as crying out for correction. Unfortunately, we also realize that most drug-based schemes for behavior modification are largely non-specific. While very desirable effects on abnormal behavior can be achieved, there are concomitant gross, more global and often deleterious effects on many normal brain functions. Systemic and untargeted pharmacological approaches, therefore, must be supplemented and eventually replaced with finer, more specific tools, some of which must be genetic. We are ethically justified, and even required, to design procedures for preventing or altering these devastating disorders and for re-establishing the normal human functions that have gone so terribly awry in these disorders.

We will certainly come to recognize and understand genetic factors correlated with desirable and with deleterious versions of behavioral traits, and the acquisition of that knowledge is important, inevitable and good. It must and will be put to good clinical use. However, once past the relatively simple examples of obvious human diseases, there are and will continue to be many instances in which the distinctions between the pathological characteristics and the many variations on the spectrum of normal human traits will be much more difficult to make and not nearly as universally accepted, either among the members of any given society or among the many different societies of the human species. Even with the use of the sophisticated tools of molecular genetics, we will not always be able to recognize or define the differences between normality and aberration. And certainly some of the most difficult instances will arise in the province of human behavior and psychiatric disease, in the expressions of the human mind: traits of intelligence, cognitive skills, aspects of personality, creativity, the ability to learn, sociability.

What is in question is how we will use all of our exploding new genetic information – for the amelioration of human suffering, or for the construction of biologically irrelevant and possibly socially dangerous barriers between classes of people, the genetic "haves" and "have nots." In all likelihood, as with all knowledge, we will do both. Therein lie some of the questions that will come to complicate many of the imminent advances in the broad field of molecular medicine and specifically in neuropsychiatry. What use will we make of the intimate new genetic knowledge that we will develop about each other? How will we use powerful detection tools when we have no correspondingly effective methods for prevention or treatment? What will the response be to the pre-clinical detection of a gene associated with a neuropsychiatric disease? In what ways will diagnostic information derived from a family linkage study for the diagnosis of an indivudual be used for other, possibly uninvolved, members of the diagnostic group? How do we ensure confidentiality of such sensitive information? How do we ensure that issues of employment and insurability and many other career issues are not affected unfairly or unjustly by information obtained by genetic screening or monitoring programs, whether they be voluntary or compulsory? Is there a justification for compulsory detection programs?

These problems must not in any way be taken as an indictment of the information itself or as pressure to inhibit the acquisition of knowledge of the biological basis for human behavioral traits, normal or pathological. The advances are certainly not free of difficulties and they will not come without a certain social cost. We know all too well from history that individuals or classes of people among us who display unusual, "deviant" or socially "unacceptable" or "undesirable" behavior have at times been viewed to be suffering from mental disorders, and have been for stigmatization and even more malevolent abuse. In this century alone we have seen a number of instances in which fragments of genetic information, or rather, pseudo-information, have been used to establish unenlightened and even evil social policies. The birth and nurturing early in this century of eugenics programs and their discriminatory consequences in Great Britain and the United States (Kevles 1985), as in some of the misguided immigration policies and compulsory sterilization programs of retarded patients in some States (endorsed by the U.S. Supreme Court in 1927 with the ringing phrase, "Three generations of imbeciles are enough;" Buck v. Bell, U.S. Supreme Court, 1927), the missteps in the establishment of broad screening programs for the putative antisocial XYY syndrome and for sickle cell anemia and, of course, the instituionalization of social dissidents, the Nazi experience, all need not recounting here. In each case, at least part of the justification for misguided and malevolent social and state programs was based on poor but somehow pervertedly popular genetic science, such as the establishment by the respected Cold Spring Harbor Eugenics Record Office of the genetic causes for shiftlessness, nomadism and love of the sea ("thalassophilia"). Furthemore, it will not escape attention that many of these programs have been aimed at questions close to the heart of psychiatry – the judgment by a society of the behavior of its citizens, the struggle between the free will of autonomous human persons and the more biologically determined behaviors, behaviors that may, therefore, be largely beyond human control. They touch on the relationship between those with "flawed" personalities and the state, on questions of treatment, ostracism and punishment for deviant behavior.

Despite these lamentable parts of our genetic history, they are not entirely or even largely abuses engendered principally *by* Science; rather, they are abuses *of* Science by other sociopolitical institutions. We should and must proceed with work aimed at an understanding of the interaction of the genetic information of human genes with development and behavior. To suggest the opposite is to deny to many millions of sufferers the potentially enormous benefits of a truly rational understanding of human biology and a truly rational treatment of disease. It is, of course, clear that some socially aberrant or deviant behavior is indeed pathological. But it seems very likely that much of it is not. The current and still emerging techniques of molecular genetics will certainly begin to help clarify some of the distinctions between normal and pathological behavior, the biological factors that influence a susceptibility to psychiatric disease and the mechanisms of the environmental-biological interactions that produce psychiatric disease. But we will for a long time be left with many instances in which the balance of environmental and genetic factors will remain poorly defined.

One of the major reasons for the present enthusiasm in the biomedical community for a molecular genetic approach to neuropsychiatric disease is the realization that most tools available today for the diagnosis, prevention and treatment of neuropsychiatric disease are simply not adequate. The present emphasis on genetic and biological mechanisms is in response to an enormous need to develop more effective diagnostic and therapeutic techniques. The tools of molecular genetics will and must be used to detect, prevent and treat psychiatric disease. But in proceeding, we should not lose sight of some difficult ethical and public policy issues inherent in the acquisition and use of intimate personal and possibly threatening genetic information. We must also keep in mind the particularly sensitive nature of genetic information and its uniquely sensitive relationship to neuropsychiatric disorders. We must ensure that the diagnostic and therapeutic technologies do more than open a Pandora's box of dangerous or uncontrollable technology. By interacting with institutions such as the European Science Foundation and the National Institutes of Health in the United States among others, agencies such as the World Health Organization can play an important role in identifying the genetic components of psychiatric diseases, in identifying resources and materials to study cultural and cross-cultural aspects of behavior and in establishing guidelines for the ethical use of genetic information for the relief of human disease and suffering. It is important that WHO and similar organizations organize and support more workshops and meetings to catalyze the discussion of the ethical impact of molecular genetics on psychiatry.

References

Childs B (1970) Sir Archibald Garrod's conception of chemical individuality: a modern appreciation. N Engl J Med 282:71–77

Friedmann T (1989) Progress toward human gene therapy. Science 244:1275–1281

Friedmann T (1990) The human genome project – some implications of extensive "reverse genetic" medicine. Am J Hum Genet 46:407–414

Gage FH, Wolff JA, Rosenberg MB, Xu L, Yee J-K, Shults C, Friedmann T (1987) Grafting genetically modified cells to the brain: possibilities for the future. Neuroscience 23:795–807

Gusella JF, Wexler NS, Conneally PM, Naylor SL, Anderson MA, Tanzi RE, Watkins PC, Ottina K, Wallace MW, Sakaguchi AY, Young AB, Shoulson I, Bonilla E, Martin JB (1983) A polymorphic DNA marker genetically linked to Huntington's disease. Nature 306:234–238

Kevles DJ (1985) In the name of eugenics. Knopf, New York

Riordan JR, Rommens JM, Kerem B-S, Alon N, Rozmahel R, Grszelczak Z, Zielenski J, Lok S, Plavsic N, Chou JL, Drumm ML, Iannuzzu MC, Collins FS, Tsui L-C (1989) Identification of the cystic fibrosis gene: cloning and characterization of complementary DNA. Science 245:1066–1073

Rosenberg MB, Friedmann T, Robertson RC, Tuszynski M, Wolff JA, Breakefield XO, Gage FA (1988) Grafting genetically modified cells to the damaged brain: restorative effects of NGF expression. Science 242:1575–1578

Sherrington R, Brynjolfsson J, Petursson H, Potter M, Dudleston K, Barraclough B, Wasmuth J, Dobbs M, Gurling H (1988) Localization of a susceptibility locus for schizophrenia on chromosome 5. Nature 336:164–167

St George-Hyslop PH, Tanzi RE, Polinsky RJ, Haines JL, Nee L, Watkins PC, Meyers RH, Feldman RG, Pollen D, Drachman D, Growdon J, Bruni A, Foncin JF, Salmon D, Frommelt P, Amaducci L, Sorbi S, Piacentini S, Stewart GD, Hobbs WJ, Conneally PM, Gusella JF (1987) The genetic defect causing familial Alzheimer's disease maps on chromosome 21. Science 235:885–890

Wolff JA, Fisher LJ, Xu L, Jinnah HA, Langlais PJ, Iuovone PM, O'Malley KL, Rosenberg MB, Shimohama S, Friedmann T, Gage FH (1989) Grafting fibroblasts genetically modified to produce L-dopa in a rat model of Parkinson disease. Proc Natl Acad Sci USA 86:9011–9014

Ethical Aspects of Human Genome Mapping and Sequencing

V. Pačes

Present methods of molecular cloning and DNA sequencing are so advanced that the determination of the complete nucleotide sequence of the human genome is nearly possible (Cantor 1990; Watson 1990). Several meetings and discussions have been devoted to strategies suitable for such a huge project, to technical problems associated with it and to plans about how to organize it. Less attention has been aimed at ethical problems and security aspects of human genome mapping and sequencing. The information obtained from such a project would be overwhelming, and we can certainly think of a number of ways it could be used and misused.

The haploid human genome consists of 22 autosomes and two heterochromosomes containing approximately 3×10^9 base pairs. It is estimated that only about 5 % to 10 % of the genome encodes molecules of RNAs and proteins. The remaining DNA represents nucleotide sequences regulating DNA replication and transcription and sequences of unknown function. These last mentioned DNA and sequences are interesting; they were maintained during evolution and some of them are arranged into arrays of repetitive sequences, some of which underwent convergent evolution. The function of these sequences, although mysterious, may be important. However, they also may represent either evolutionary fossils without any direct importance for present life forms, or so-called "selfish DNA", the function of which is to perpetuate itself. This division of human DNA (and DNA of other species) into a clearly functional portion and a large portion without function, or with unknown function, poses the question of whether it would not be wasteful to map and sequence the whole genome. For instance, the sequencing of cDNAs might may considerably promote our knowledge of those genetic diseases which are caused by mutations of particular genes. If cDNA derived from mRNA isolated from various tissues were sequenced, differences of expression of certain genes in various organs, tissues and cells would be detected. If such an approach were applied to various species, we might see the differences between genes of those species. However, genome sequencing rather than cDNA sequencing would give us complete information about man. We would know the nucleotide sequences of all genes regardless of how and whether they are expressed, and we would obtain a complete account of regulatory sequences, and sequences of unknown function. This might eventually lead to an explanation of their functions and the mechanisms of these functions.

Srám, Bulyzhenkov, Prilipko, Christen (Eds.)
Ethical Issues of Molecular Genetics in Psychiatry

It seems that most scientists consider full genome mapping and sequencing the better alternative. With current methods of DNA sequencing, a small team of, say, two skilled specialists is able to determine reliably a sequence of about 10^3 base pairs a day, which adds upto 3×10^5 base pairs a year. If 100 laboratories were involved in the project, they would be able to sequence 3×10^7 base pairs a year.

This calculation may underestimate the sequencing efficiency of good laboratories, but it does not take into account large DNA fragments isolation, their digestions and cloning and recloning in specialized vectors. These operations themselves are not without problems, and may consume considerable time.

Thus it is evident that, with available techniques, the sequencing of the complete human genome is difficult to organize. Instead, any effort should concentrate on improvement of cloning and sequencing techniques, to speed up the final output of nucleotide sequences by at least ten times.

The first project to map and sequence the human genome was started in the United States by the Department of Energy (DOE). The National Institutes of Health (NIH), together with the DOE, now supports American laboratories developing the new methods and techniques necessary to increase DNA sequencing efficiency (Barnhart 1990). It is necessary to develop new cloning strategies by which large DNA fragments can be isolated and amplified. New vectors are constructed for this purpose, namely cosmids and yeast artificial chromosomes (YAC). Cosmids are plasmids with cos sites of bacteriophage lambda, and are used for efficient introduction of recombinant DNA into cells of *Escherichia coli* by bacteriophage-lambda infection. Although DNA fragments up to 50 kilobase pairs can be cloned, purified and amplified in cosmids, many millions of clones would be neccessary to cover a single human chromosome. A more promising vector is YAC, consisting of CEN and ARS loci (ensuring faithful replication and segregation of the vector), markers and a polylinker. DNA fragments up to 700 kilobase pairs can be cloned in this type of vector (Burke et al. 1987). Large DNA fragments needed for such cloning can be obtained by pulsed-field gel electrophoresis (Barlow and Lehrach 1987) of DNA digested by restriction enzymes recognizing rare nucleotide sequences (e.g. NotI).

Optimization of cloning strategies should be accompanied by improved methods of clone handling, arranging of DNA fragments into contigs and creating specialized software (Bell and Marr 1990).

For large DNA sequencing projects, Sanger's dideoxy method is often used (Sanger et al. 1977). The final stages of the method, i.e. electrophoresis, sequence reading and data storing and handling can be automated, and several instruments are available commercially, using either radioactive or fluorescent labels. Many specialists feel that an improvement of the sequencing procedure itself can hardly be substantial at the level of electrophoretic resolution of oligonucleotides. Alternative techniques, e.g., scanning tunneling microscopy, are now being investigated.

Let us now turn our attention to the ethical aspects of human genome mapping and sequencing. There are two basic questions that should be answered:

1. Is it ethical to determine the full nucleotide sequence of human DNA? Have we the right to know our own hereditary message?
2. If so, what should be done with the information we obtain? Should it be available to everybody? Might it serve as the basis for future genome manipulations – and what kind of manipulations?

The answer to the first question is probably yes, we have the right to read our own genetic information. The knowledge itself is not harmful and it can be used in many ways that are profitable to human beings. However, is now the right time to do so? Many scientists point out that such a big project may cost too much and that funding of other, even more intellectually challenging scientific projects will be cut down. This may apply even to important projects in biomedical research, if the human genome sequencing project becomes too prestigeous, and gets high priority. An argument that is sometimes heard is that the knowledge of the human DNA will automatically solve many problems which otherwise would have to be solved separately in costly projects. There is no doubt that information obtained from the human nucleotide sequence will promote many areas of biological science. Although full consensus on these issues has not been reached, DOE and NIH have started to finance the human genome project as described above, and it seems that the United States is taking on this ambitious project to demonstrate its superiority in biological research. The less rich and smaller countries, which will not participate in the human genome project, will probably be considered inferior. And here I see a danger that these countries will allocate some money for the human genome sequencing project simply for this reason, and money will then be lacking in other important areas of research. A better position for some countries may be not to take part in the human genome sequencing project, and to spend money so saved in other projects where they may then dominate or at least significantly contribute. In view of the sums the United States plans to spend on the human genome sequencing project, any significant contribution by these countries is highly unlikely.

It is now clear that deciphering of the human genetic message is too ambitious a project not to be solved, especially in view of its technical feasibility. However, it might be difficult to find enough scientists willing to work on the project full time. They may be reluctant to participate in what they may consider tedious and not very inspiring work. Although many scientists may this way it is likely that they would be interested in knowing the sequence, interpreting it and working with information derived from it.

This brings us to questions that are more difficult to answer. What should be done with the sequence? What should be permitted to be done with the information obtained from the sequence? Should it be available to everybody?

Although the sequence itself will certainly yield interesting information, the more important results will come from comparison of the sequence of one individual with the sequence of other human individuals. This would give us a notion of diversity, and of tolerated variations of the human genome. It would also make it easier to detect genetic defects. Knowledge of the nucleotide sequence of genomes of several individuals will lead to a better understanding of evolutionary events at the molecular level, of the mechanisms by which mutations are introduced into hereditary material or eliminated from it or fixed in it, and of the nature of genetic diseases. All this information will help us to understand man better, and to improve his life.

Equally interesting and important would be to compare, at the molecular level the human genome with genomes of other animals. If such a comparison is made between man and other primates closely related to man, the results may help to answer the fundamental question of what the basis of difference is between man and other animals. Which genes or mechanisms are responsible for the ability of man to abstract? Which genes are responsible for the sudden increase of brain volume in man compared to other animals?

From the technical point of view, such comparisons will of course be difficult to perform in the next few decades without an essential breakthrough in cloning and sequencing technologies, since they could only be done after the complete nucleotide sequences of all genomes in question are determined.

Ethical objections to such studies will probably not be too serious. More controversial, however, may be the problem of who should be eligible to obtain specific information about a person's DNA sequence. In particular, defective genes may be objects for misuse. However, this information cannot be obtained from the human genome mapping and sequencing project since the person whose DNA will be sequenced will remain unknown. It is possible that the sequence will reveal completely new phenomena which may become targets for misuse, not against a particular person but against mankind. Unless human genome mapping and sequencing is organized as a secret project, it is virtually impossible not to have the results available by everybody. Especially sensitive might for instance become a comparison of nucleotide sequences of genomes derived from individuals of different races or nationalities. We might foresee a number of other potential misuses, e.g. of nucleotide sequences sensitive to integration of oncogenes or viruses. Slight changes in particular genes or their regulatory sequences might have unpredictable effects on the whole organism. Even without direct changes of the nucleotide sequences of a genome, there are methods available today by which expression of a genetic message can be affected, e.g., by association with antisense RNA. All possible targets for misuse should be anticipated and thoroughly discussed before any serious sequencing of the human genome begins, and clear-cut legislation should be worked out.

References

Barlow DP, Lerhach H (1987) Genetics by gel electrophoresis: the impact of pulsed field gel electrophoresis on mammalian genetics. Trends Genet 3:167–171

Barnhard BJ (1990) Human genome, 1989–90 program report. US Department of Energy, Washington

Bell GI, Marr TG (1990) Computers and DNA. Addison-Wesley, Redwood City

Burke DT, Carle G, Olsen M (1987) Cloning of large segments of exogenous DNA into yeast by means of artificial chromosome vectors. Science 236:806–812

Cantor CR (1990) Orchestrating the human genome project. Science 248:49–51

Sanger F, Nicklen S, Coulson AR (1977) DNA sequencing with chain-termination inhibitors. Proc Natl Acad Sci US 74:5463–5467

Watson JD (1990) The human genome project: past, present, and future. Science 248:44–48

Transcultural Problems in the Use of Medical Genetics in Clinical Practice

A. Serra

The appraisal of the heavy genetic load in human populations and the fear of its increase prompted Nobel Laureate Muller's appeal (1967) to human geneticists convened at the Third International Congress of Human Genetics in Chicago, to engage in a strong offensive for the control of human evolution. Some of his words deserve mention: "Modern culture by maximal saving of lives and fertility, unaccompanied by a conscious planning which takes the genetic effects of this policy into account, must protect mutations detrimental to bodily vigor, intelligence or social predisposition [...]. If genetic defects and shortcomings were to be allowed to accumulate to an unlimited extent among us, as seems to be happening now, the condition would eventually be reached in which each person likewise would present an immense, yet in his case distinctive, complex of problems of diagnosis and treatment" (pp. 526, 527). These statements were not exaggerations. A few years later, in a Technical Report of the World Health Organization (1972), a group of experts estimated that, if in the next generations all the subjects who up to now have not reached the age of fertility would reproduce, then the incidence of polygenic diseases should increase 3–5 % per generation, and the incidence of monogenic diseases should increase 15 % per generation globally, and would double after seven generations. Muller's appeal and proposals were echoed and supported by other distinguished geneticists (Sonneborn 1965; Lederberg 1966). Medicine, which at that time was being attracted by the power of the rapidly growing field of human genetics to promote a better understanding of human disease, felt inevitably involved in that enterprise (Motulsky 1968), and Medical Genetics developed as a new branch. From that time on many important and beautiful pages in the history of Medicine have been written as a result of the close collaboration between the clinics and the newly developing areas of genetics: cytogenetics, immunogenetics, biochemical and mathematical genetics, to mention only a few. In this paper I shall first sketch very briefly how new scientific achievements have passed, through medical genetics, into clinical practice. Secondly, I shall attempt to indicate the main causes of the ethical tension that characterizes this new era of Medicine, and actually represents the fundamental transcultural problem in the application of medical genetics in clinical practice.

In the 1960s and 1970s an extraordinary number of new genetic diseases and congenital anomalies were discovered to be due to either chromosomal aberrations (Borgaonkar 1980; de Grouchy and Turleau 1982) or to single gene

Srám, Bulyzhenkov, Prilipko, Christen (Eds.)
Ethical Issues of Molecular Genetics in Psychiatry

mutations, with the number of the latter increasing from 1487 in 1966 to 2811 in 1978 (McKusick 1988). Most important, with the use of new biochemical techniques it was possible to demonstrate, in a number of these genetic diseases, either a reduced amount of certain metabolites, or their accumulation and excessive excretion, or a deficit of functionally significant protein, suggesting that enzyme or specific protein defects were the primary effect of gene alteration (Harris 1975; Galjaard 1980). By 1980 the responsible molecular defect was elucidated for about 200 genetic diseases, a great step forward although this represented less than 10 % of the conditions proven or suspected to be due to single genes. Furthermore, for many of them not only the etiology but also the pathogenesis could be studied extensively (Galjaard 1980; Stanbury et al. 1983), and valuable genetic parameters, like gene frequency, heterozygote incidence in the populations and segregation patterns were established (Collaborative Study 1975). Attempts to identify heterozygotes for deleterious genes were also made, but they frequently proved unsuccessful, due to ample value overlappings between controls and carriers, or unjustified, because of their rarity, or even impossible, because of a lack of reliable detection methods. Any kind of effective treatment and cure was also impossible. Yet, remarkable improvements in ultrasonography (Donald 1969; Jouppila 1971; Hansman 1979) and in the techniques of amniocentesis (Scrimgeour 1973) and fetoscopy (Rauskolb 1979; Scrimgeour 1979) paved the way for the introduction and diffusion of prenatal diagnosis in medical practice. By the end of 1977 summaries of the two largest surveys on prenatal diagnosis of pregnancies at risk reported from the USA and Canada (13,785; Epstein and Golbus 1978) and from Western Europe and Israel (11,334; Murken et al. 1979; Galjaard 1980) showed that the number of fetuses found to be affected by genetic diseases was approximately 4 %.

In the 1980s the explosion of new technologies (Donis-Keller and Botstein 1988) that followed the brilliant discovery of recombinant DNA by the Nobel Laureate Berg (1981) caused the molecular revolution in medical genetics that altered altogether the direction of the search: the major interest was no longer the defective product of the genes but the genes themselves, looking for their precise errors and their direct effects. The 127 papers presented at Cold Spring Harbor in 1986 under the title of "Molecular Biology of Homo Sapiens," covering 1229 pages (Cold Spring Harbor Symposia on Quantitative Biology 1986), gives an exciting idea of the successes of this new era of human and medical genetics. In less than ten years, more than 450 disease genes (Harper et al. 1989), including oncogenes and anti-oncogenes (Varmus 1984; Hall 1986; Knudson 1986; Bradshaw and Prentis 1987; Klein 1987) and approximately 3300 anonymous DNA sequences (McAlpine et al. 1989), were mapped in human chromosomes. The latter were shown to be extremely useful in locating coding genes through linkage studies (Donis-Keller et al. 1987; White and Lalouel 1988). A tedious but highly rewarding molecular analysis of a few mutated genes, such as those of beta-hemoglobinopaties and hemophilia (Kazazian and Antonarakis 1988), Duchenne and Becker type muscular dystrophy (Burgess et al. 1987; den Dunnen et al. 1987; Forrest et al. 1987) and cystic fibrosis (Kerem et al. 1989; Riordan et al. 1989; Rommens et al. 1989), disclosed not only the great heterogeneity of mutations that can be responsible for a given disease, but also the possibility of discovering the specific product of the normal gene, thus facilitating the understanding of the pathogenesis of single gene diseases. As a consequence of these advances, the possibility of identifying patients, actually or possibly affected by genetic disease, and heterozygous carriers increased enormously, and became relatively easier through both postnatal and prenatal diagnosis (Donis-Keller et al. 1986; Schmidtke 1986; Caskey 1987; Martin 1987; Antonarakis 1988; Boehm 1988; Darras et al. 1988; Phillips 1988), which was made available very precociously by the introduction of chorionic villi sampling (Kazy et al. 1982; Ward et al. 1983; Simoni et al. 1984; Goldberger and Golbus 1988; Brambati et al. 1990).

The enthusiasm of medical geneticists went still further. Could not the acquired ability of manipulating genes eventually be used to cure genetic disease? Tissue culture experiments had shown that purified genes could be transferred into mammalian cells (bone marrow progenitor cells, lymphocytes, fibroblasts) and function sufficiently well to alter the phenotype of the cells (Bacchetti and Graham 1977; Miller et al. 1986; Parkman 1986) and even cure them, at least partially, as in the case of lymphoblasts from patients affected by Lesch-Nyhan disease (Willis et al. 1984; Nelson et al. 1986), of fibroblasts from patients affected by Gaucher disease (Sorge et al. 1986), and of fibroblasts or bone marrow cells from transgenic laboratory mammals affected by phenylketonuria (Ledley et al. 1985) or adenine deaminase deficiency (Belmont and Enkel-Tigges 1986). Thus, the perspectives for somatic cell gene therapy first delineated by Anderson in 1984 appeared attainable. Further investigations aimed at making the transfection more regulated, and the risk of insertional mutations minimal (Gregg and Smithies 1986; Thomas and Capecchi 1986; Selden et al. 1987; Dzierzak et al. 1988), could finally substantiate the proposal for the first projects of human gene therapy that were submitted to competent license boards (Orkin and Williams 1988).

Simultaneous to these developments, refinement of gene transfer, including direct microinjection of genes into cell nuclei (Capecchi 1980), and progress in embryology permitted a major advance toward the goal of gene therapy (Gordon 1990). It was demonstrated that cloned DNA could be found in the cells of newborn mice when microinjected into the pronuclei of fertilized mouse eggs, and that the newly acquired genes were transmitted to the germ line of these transgenic mice. Subsequently their quantitative and tissue-specific expression appeared controllable, and accurately designed experiments explored, in mice, the potentials of gene microinjection in germ cells for gene therapy. Hammer et al. (1984) obtained a partial correction of murine hereditary growth disorder, Costantini et al. (1986) successfully treated mice with beta-thalassemia by transfer of human adult beta-globin gene, and Mason et al. (1986) restored reproductive functions in hypogonadal mice, which were sterile because of the absence of normal genes at the locus encoding for the gonadotrophin-releasing hormone (GnRH), by the microinjection of the cloned wild-type GnRH gene.

However, a rigorous analysis of the results obtained so far in experimental animal models shows that the prerequisites for good success – that is, that gene transfer be harmless, efficient, directed and certain to result in physiologically significant expression – are all far from being satisfied. Therefore Gordon (1990) suggests, in accordance with common consensus, that it will be many years before germ cell gene therapy becomes available.

Other techniques are now being developed that aim at preventing implantation or even the conception of affected embryos. In vitro produced embryos can be split, with one half frozen and the other half used to establish cell lines whose genetic abnormality can be evaluated with recombinant DNA techniques. If the cell lines are found to be abnormal, the frozen half would not be implanted. Furthermore, the new technique of primer-mediated enzymatic amplification (PCR) of specific sequences may allow the detection of a given gene either in single blastomeres, thus avoiding the need for freezing embryos, or even in oocytes and their first polar bodies, thus avoiding the fertilization of the carrier oocyte. With the application of these techniques not only sex identification of preimplantation mouse and human embryos (Gordon 1990; Handyside et al. 1990), but also preimplantation disease diagnosis in mouse models (Monk et al. 1987; Holding and Monk 1989), and carrier oocyte detection in humans (Monk and Holding 1990) have already been performed. Further refinements are still needed, but their introduction into medical practice would eliminate "the necessity of abortion," according to Gordon (1990) and, according to Monk and Holding (1990), might "circumvent some of the ethical difficulties of dealing with human embryos."

No one should underestimate these advances of Medical Genetics and misjudge their actual or attempted application in the clinical practice. However, it must be recognized that many of these advancements occurred in an atmosphere of ethical tension. On one side, incorrect or incomplete information and/or emotional excitement ("genicity" to use Capron's neologism) led to an irrational condemnation or indiscriminate ban on new scientific approaches to very serious human problems. The issues raised, especially those regarding genetic engineering (Rifkin 1983), fortunately urged public debate that, in the most controversial cases, was channeled into politically responsible institutions (Murray 1990) which were, and still are, trying to establish a reasonable and effective public policy. On the other side, well-informed humanistic groups and scholars (McCormick 1981; World Council of Churches 1982; Motulsky 1983), and even a number of the most enthusiastic leaders in the development and application of the new technologies of medical genetics (Anderson 1989), have deep-rooted concerns about the possibility of misuses, and want clear lines to be drawn between what is permissible and what is not. Obviously these attitudes are merely an expression of the tension that still surrounds and crosses many new areas of investigation and application of medical genetics. This tension is felt as a disturbing force by the majority of those directly engaged in the field. Understanding the hidden cause of that tension is highly important in order to foresee its possible effects and implications. The fundamental hidden cause is, I believe, the difference in the cultural background upon which individuals or societal groups base the ethical principles of behavior. I shall briefly illustrate this in three main areas where medical genetics is actively engaged in clinical practice.

1. In the first section of this chapter we saw that there is a rapidly increasing possibility of prenatal diagnosis of genetic disease. Negative eugenics, through selective termination of affected fetuses, is at present the preferred policy, but is not without ethical tension (Dunstan 1988; Annas and Elias 1990). It is easy to see how, due to different cultural foundations, some people may regard selective termination as ethically good, some people may see it as ethically good under certain circumstances, and others may see it as ethically wrong in any case. Aiken (1973), of the Department of Philosophy of Brandeis University at Waltham (Mass.), stated that "claims regarding the right to biological survival are entirely contingent upon the ability of the individual in question to make, with the help of others, a human life for himself. This means that in circumstances where there exists no possibility of anything approaching a truly human life, the right to biological or physical survival loses its own raison d'etre and hence that the merciful termination of life, in the bio-physical sense is acceptable or perhaps even obligatory" (p. 180). Given this principle, according to which the right to survival of a conceptus is linked to the possibility of a vaguely defined "truly human life", the selective termination of at least the large majority of genetically affected fetuses is ethically good.

In its final report, Consultation on Genetics and Quality of Life (Birch and Abrecht 1975), World Council of Churches declared that "a decision for foetal diagnosis and abortion is a weighty decision, as the foetus, although still dependent, has a potential existence as an independent human being. The decision to deprive it of that potentiality depends on a conclusion that the detriments resulting from its birth outweigh the benefits" (p. 207). The principle here is certainly more restrictive, since the basic criterion is the human being itself, but the conditional criterion that selective termination of a fetus becomes ethically good when disadvantages from its birth outweigh the benefits practically cancels any limitation, thus opening the door to unlimited relaxation. For instance, what only a few years ago was considered wrong from any ethical standpoint – such as the selective termination of a Klinefelter or Turner fetus or of a fetus of unwanted sex – is now becoming routine, and there are pressures to extend selective termination to heterozygous carriers of defective genes (Fuchs 1972; Hsia 1972).

A third ethical view was recently condensed in the following declarations of an Instruction of the Catholic Church (Congregation for the Doctrine of the Faith 1987): "The fruit of human generation from the first moment of its existence, that is to say from the moment the zygote is formed, demands the unconditioned respect that is morally due to the human being in his bodily and spiritual totality.... From that same moment his rights as a person must be recognized, among which in the first place is the inviolable right of every innocent human being to life" (p. 13). On the basis of this unambiguous principle, the selective termination of any affected fetus is ethically wrong.

It is not my task here to critically examine the above-mentioned principles, which are comprehensive of many variants traceable in other cultures. My intention is only to demonstrate that the ethical tension pervading the area of medical genetics involved in the prevention of genetic disease has its deep roots in the different culturally transmitted understanding of what "truly human life," "human subject," "human dignity" and "human rights" are, with all possible historical changes.

2. More subtle, yet less dramatic, is the situation of the ethical tension relative to carrier detection and preventive counseling. The ongoing debate on population screening for carriers of the cystic fibrosis gene – approximately 1 in 25 Caucasians (Roberts 1990; Colten 1990; Editorial 1990; Super et al. 1990; Gilbert 1990) – and the published position of the American Society of Human Genetics (Caskey et al. 1990) that "routine CF carrier testing of pregnant women and other individuals is NOT yet the standard of care in medical practice" are an indication of the conflict of opinions on the subject. This conflict is even larger and stronger in the field of psychiatric diseases (Bloch and Hayden 1990; Murray and Gill 1990; Wertz 1990).

In this context, three group of individuals are to be considered: the carrier, the carrier's relatives and the society at large. Supposing that all the technical problems are solved and that the proper genetic condition of the gene (dominant, recessive or X-linked) is taken into account for a correct screening design, many questions about "who" should or could be tested and "when", about "who must advise" prospective candidates before and after testing, about the needed services and about possible reactions on the part of the candidates themselves and of social groups (employers, insurances, justice) towards the carriers should be examined and answered before starting any project (President's Commission 1983; Harper and Clarke 1990; Friedmann 1990). But even in the answers to these questions the impact of the cultural background is evident. To some the individual's autonomy, privacy and interest must prevail over the interests of relatives or society. To others, the opposite should prevail: relatives at risk must be protected, safe workplaces and productive workers should be ensured, and public health should be protected and promoted for future generations.

Reconciling these different views, however, could be easier than in the case of genotype selection. This is particularly true if each intended screening program is considered separately when it is to be applied both to the population and/or single families.

3. Gene therapy is a third area where differences in cultural background are the source of vivid ethical tensions. In tracing the evolution of ethical debate about human gene therapy, Fletcher (1990), a moral theologian, underlines, on one end, the large ethical consensus now reached for somatic cell gene therapy. On the other end, discussing the ethics of experimenting with germ cell gene therapy, he remarks that "to argue that it should never be done at all is unreasonable at the least, examined from the stand point of the well-being of future generations. One generation can learn if germ-cell therapy is feasible. In their own time, other generations can decide if and how to use this knowledge" (p. 66). Anderson (1985, 1989), the pioneer of gene therapy, is absolutely convinced that somatic cell gene therapy has a great potential for reducing or eliminating the suffering accompanying genetic diseases, and that it is ethically good. And the majority of those working in the field agree with him (Ledley 1987; Friedmann 1989). However, he is concerned that this first step opens the door to the two next logical steps: 1) germ line gene therapy; and 2) somatic cell

and germ line engineering, that could allow intentional alteration of the human genome. This last step, according to him, would detract from the dignity of man, and therefore would be morally wrong, in that it threatens important human values in two ways: first, because harm could be caused, with the risks exceeding potential benefits; and second, because it would be morally precarious, as it is impossible to determine what genes should or could be provided, who should or could receive them, and how to prevent discrimination against those who do or do not receive a gene. Therefore, based on these reflections, he proposes that "on medical and ethical grounds a line should be drawn excluding any form of enhancement engineering" (p. 689). But, he adds, "as experience is gained, the line should be moved to include possibly germ line gene therapy for specific diseases" (p. 690).

However, germ line gene therapy and all of the experimentation that should obviously precede it, if conducted or attempted in human embryos conceived in vitro (which seems to be the only practical way to operate), is considered ethically unacceptable by all those who view the production of human embryos in vitro, for whatever reason, as a violation of human dignity (Congregation for the Doctrine of the Faith 1987).

It would be out of place to enter here into the details of the multiple approaches to gene therapy that could and will slowly be introduced in clinical practice, or to discuss which of the different ethical positions is preferable. However, it is clear that the different positions are founded in different cultural backgrounds.

This very brief and limited analysis of the hidden source of the ethical tension that permeates the most advanced frontiers of medical genetics suggests that this tension will not easily subside.

All the efforts to find rules, to establish deontological codes, and to dictate laws in these areas might be helpful in preventing gross abuses, but not to release tension. It is my impression and conviction that a profound sense of human responsibility – of scientists in science-making and of physicians and society in the application of new scientific developments – may be the best guarantee against abuses, and the best tension releaser. One aspect of this sense of responsibility is, in this context, to make any effort to reduce suffering with all means afforded by a tireless search of effective treatments and humanly acceptable ways of prevention. The other aspect is to strive, as in any field of research, to make of the inevitable plurality of opinions not a matter of tension but a spur to maturation in a reciprocal constructive understanding. This sense of responsibility will probably and gradually lead to the lessening of the ethical tension, notwithstanding the persisting differences in the cultural backgrounds.

It is highly probable that Mendel did not foresee the extraordinary developments of the ingenious hypothesis he elaborated in these Moravian surroundings, or the problems raised by their inevitable use. Even if he had foreseen all this, using the stimulus of his creativity and insight he would have continued to work facing, with high responsibility, all of these problems, both scientific and human, recalling what he wrote in his second letter to Nägeli: "My happiness shall be doubled if I succeed in finding their solution."

References

Aken HD (1973) Life and right to life. In: Holton B, Callahan D, Harris M, Condliffe P, Berkley B (eds) Ethical issues in human genetics. Plenum, New York, pp 173–183

Anderson WF (1984) Prospects for human gene therapy. Science 226: 401–409

Anderson WF (1985) Human gene therapy: scientific and ethical considerations. J Med Philos 10: 275–291

Anderson WF (1989) Human gene therapy: why draw a line? J Med Philos 14: 681–693

Annas GJ, Elias S (1990) Legal and ethical implications of fetal diagnosis and gene therapy. Am J Med Genet 35: 215–218

Antonarakis SE (1988) Diagnosis of genetic disorders at the DNA level. N Engl J Med 320: 153–163

Bacchetti S, Graham FL (1977) Transfer of the gene for thymidine kinase to thymidine kinase deficient human cells by means of purified Herpes simplex viral DNA. Proc Natl Acad Sci USA 74: 1590–1594

Belmont JW, Enkel-Tigges J (1986) Expression of human adenosine deaminase in murine haematopoietic progenitor cells following retroviral transfer. Nature 322: 385–387

Berg P (1981) Dissection and reconstruction of genes and chromosomes. Science 213: 296–303

Birch C, Abrecht P (eds) (1975) On behalf of World Council of Churches: genetics and the quality of life. Pergamon, London, pp 200–223

Bloch M, Hayden MR (1990) Opinion: predictive testing for Huntington disease in childhood: challenges and implications. Am J Hum Genet 46: 1–4

Boehm CD (1988) Prenatal diagnosis and carrier detection by DNA analysis. Progr Med Genet [New Ser] 7: 143–179

Borgaonkar DS (1980) Chromosomal variation in man. A catalog of chromosome variants and anomalies, 3rd edn Liss, New York

Bradshaw RA, Prentis S (eds) (1987) Oncogenes and growth factors. Elsevier, Amsterdam

Brambati B, Lanzani A, Tului L (1990) Transabdominal and transcervical chorionic villus sampling: efficiency and risk evaluation of 2411 cases. Am J Med Genet 35: 160–164

Burgess AHM, Logan C, Hu X, Belfall B, Worton RG, Ray PN (1987) A cDNA clone from the Duchenne-Becker muscular dystrophy gene. Nature 328: 434–437

Capecchi MR (1980) High efficiency transformation by direct microinjection of DNA into cultured mammalian cells. Cell 22: 479–488

Capecchi MR (1989) Altering the genome by homologous recombination. Science 244: 1288–1292

Caskey CT (1987) Disease diagnosis by recombinant DNA methods. Science 236: 1223–1228

Caskey CT, Kaback MM, Beaudet AL, Cavalli-Sforza LL (1990) The American Society of Human Genetics statement on cystic fibrosis screening. Am J Hum Genet 46: 393

Cold Spring Harbor Symposia on Quantitative Biology (1986) Molecular biology of Homo sapiens. Cold Spring Harbor Laboratory, Cold Spring Harbor

Collaborative Study (1975) on the frequency of inborn errors of metabolism. Humangenetik 30: 273–286

Colten HR (1990) Screening for cystic fibrosis: public policy and personal choice. N Engl J Med 322: 328–329

Congregation for the Doctrine of the Faith (1987) Instruction on respect for human life in its origin and on the dignity of procreation. Poliglotta Vaticana, Vatican City

Costantini F, Chada K, Magram J (1986) Correction of murine beta-thalassemia by gene transfer into the germ line. Science 233: 1192–1194

Darras BT, Koenig M, Kunkel LM, Francke U (1988) Direct method for prenatal diagnosis and carrier detection in Duchenne/Becker muscular dystrophy using the entire dystrophin cDNA. Am J Med Genet 29: 713–726

De Grouchy J, Turleau C (1982) Atlas des maladies chromosomiques, 2nd edn. Expansion Scientifique Française, Paris

Den Dunnen JT, Bakker E, Klein Breteler EG, Pearson PL, van Ommen G-JB (1987) Direct detection of more than 50% of the Duchenne muscular dystrophy mutations by field inversion gels. Nature 329: 640–642
Donald I (1969) Ultrasonic in diagnosis (sonar). Proc R Soc Med 62: 442–446
Donis-Keller H, Botstein D (1988) Recombinant DNA methods: application to human genetics. Progr Med Genet [New Ser] 7: 17–42
Donis-Keller H, Barker DF, et al. (1986) Highly polymorphic RFLP probes as diagnostic tools. Cold Spring Harbor Symp Quant Biol 51: 317–324
Donis-Keller H, Green P, et al. (1987) A genetic linkage map of the human genome. Cell 51: 319–337
Dunstan GR (1988) Screening for fetal and genetic abnormality: Social and ethical issues. J Med Genet 25: 290–293
Dzierzak E, Papayannopoulou T, Mullingan RC (1988) Lineage specific expression of a human beta-globin gene in murine bone marrow transplants recipients reconstituted with retrovirus-transduced stem cells. Nature 331: 35–41
Editorial (1990) Cystic fibrosis: prospects for screening and therapy. Lancet 335: 79–80
Epstein CJ, Golbus MS (1978) The prenatal diagnosis of genetic disorders. Annu Rev Med 29: 117–128
Fletcher JC (1990) Evolution of ethical debate about human gene therapy. Hum Gene Ther 1: 55–68
Forrest SM, Cross GS, Speer A, Gardner-Medwin D, Burnas J, Davies KE (1987) Preferential deletion of exons in Duchenne and Becker muscular dystrophies. Nature 329: 638–640
Friedmann T (1989) Progress toward human gene therapy. Science 244: 1275–1281
Friedmann T (1990) Opinion: the human genome project. Some implications of extensive "reverse genetic" medicine. Am J Hum Genet 46: 407–414
Fuchs F (1972) Discussion to the paper of Hsia DYY In: Harris M (ed) Ethical problems in human genetics: early diagnosis of genetic defects. DHEW, Washington
Galjaard H (1980) Genetic metabolic diseases. Early diagnosis and prenatal analysis. Elsevier/North-Holland, Amsterdam
Gilbert F (1990) Is population screening for cystic fibrosis appropriate now? Am J Hum Genet 46: 395–396
Goldberger JD, Golbus MS (1988) Chorionic villus sampling. Adv Hum Genet 17: 1–25
Gordon JW (1990) Micromanipulation of embryos and germ cells: an approach to gene therapy? Am J Med Genet 35: 206–214
Gregg RG, Smithies O (1986) Targeted modification of human chromosomal genes. Cold Spring Harbor Symp Quant Biol 51: 1093–1099
Hall A (1986) Oncogenes. Genet Eng 5: 61–116
Hammer RE, Palmiter RD, Bringster RL (1984) Partial correction of murine hereditary growth disorder by germ-line incorporation of a new gene. Nature 371: 15–67
Handyside AH, Kontogianni EH, Hardy K, Winston RML (1990) Pregnancies from biopsied human preimplantation embryos sexed by Y-specific DNA amplification. Nature 344: 768–770
Hansman M (1979) Sonar in prenatal diagnosis. In: Murken JD, Stengel-Rutkowski S, Schwinger E (eds) Prenatal diagnosis. Enke, Stuttgart, pp 155–183
Harper PS, Clarke A (1990) Should we test children for "adult" genetic disease? Lancet 335: 1205–1206
Harper PS, Frézal J, Ferguson-Smith MA, Schinzel A (1989) Report of the Committee on Clinical Disorders and Chromosome Deletion Syndromes. Human Gene Mapping 10. Cytogenet Cell Genet 51: 563–611
Harris H (1975) The principles of biochemical genetics, 2nd edn. North-Holland, Amsterdam
Holding G, Monk M (1989) Diagnosis of beta-thalassemia by DNA amplification in single blastomeres from mouse preimplantation embryos. Lancet 2; 532–535
Hsia DYY (1972) Detection of heterozygotes. In: Harris M (ed) Ethical problems in human genetics: early diagnosis of genetic defects. DHEW, Washington, pp. 105–120 (Fogarty International Center publication, proceedings no 6)
Human Gene Mapping 10 (1989) New Haven Conference. 10th International Workshop on Human Gene Mapping. Cytogenet. Cell Genet. 51, vols 1–4

Jouppila P (1971) Ultrasound in the diagnosis of early pregnancy and its complications. Acta Obst et Gynecol Scand [Suppl] 15
Kazazian HH Jr., Antonarakis SE (1988) The varieties of mutation. Progr Med Genet [New Ser] 7: 43–67
Kazy Z, Rozovsky IS, Bakharev VA (1982) Chorion biopsy in early pregnancy: a method of early prenatal diagnosis for inherited disorders. Prenatal Diagn 2: 39–45
Kerem B-S, Rommens JM, Buchanan JA, et al. (1989) Identification of the cystic fibrosis gene: genetic analysis. Science 245: 1073–1080
Klein G (1987) The approaching era of the tumor suppressor genes. Science 238: 1539–1545
Knudson AG (1986) Hereditary cancer, oncogenes and antioncogenes. Cancer Res 45: 1437–1443
Lederberg J (1966) Experimental genetics and human evolution. The American Naturalist 100: 519–525, 1966
Ledley FD (1987) Somatic gene therapy for human disease: a problem of eugenics? Trends Genet 3: 112–115
Ledley FD, Grenett HE, et al. (1985) Gene transfer and expression of human phenylalanine hydroxilase. Science 228: 77–79
Martin JB (1987) Molecular genetics: applications to the clinical neurosciences. Science 238: 765–772
Mason AJ, Pitts SL, Nikolics K, Szonyi E, Wilcox JN, Seeburg PH, Stewart TA (1986) The hypogonadal mouse: reproductive functions restored by gene therapy. Science 234: 1372–1378
McAlpine PJ, Shows TB, Bouchfix C, Stranc LC, Berent TG, Pakstis AJ, Douté RC (1989) Report of the Nomenclature Committee and the 1989 Catalog of Mapped Genes. Human Gene Mapping 10. Cytogenet. Cell Genet. 51: 13–66
McCormick RA (1981) How brave a new world? Dilemmas in bioethics. Doubleday, Garden City
McKusick VA (1988) Mendelian inheritance in Man, 8th edn. Johns Hopkins University Press, Baltimore
Miller AD, Palmer TD, Hock RA (1986) Transfer of genes into human somatic cells using retrovirus vectors. Cold Spring Harbor Symp. Quant. Biol. 51: 1013–1019
Monk M, Holding C (1990) Amplification of a beta-haemoglobin sequence in individual human oocytes and polar bodies. Lancet 335: 985–988
Monk M, Handyside A, Hardy K, Whittingam D (1987) Preimplantation diagnosis of deficiency of hypoxanthine phosphoribosyltransferase in a mouse model for Lesch-Nyhan syndrome. Lancet 2: 423–425
Motulsky AG (1968) Human genetics, society and medicine. J Hered 59: 329–342
Motulsky AG (1983) Impact of genetic manipulation on Society and Medicine. Science 219: 135–140
Muller HJ (1967) What genetic course will man steer? In: Crow JF, Neel JV (eds) Proceedings of the 3rd International Congress on Human Genetics. Johns Hopkins, Baltimore, pp 521–535
Murken JD, Stengel-Rutkovski S, Schwinger E (eds) (1979) Prenatal diagnosis Proceedings of the 3rd European conference on prenatal diagnosis of genetic disorders. Enke, Stuttgart, pp 1–14, 22–28, 34–45, 65–72, 73–84, 88–93
Murray RM, Gill M (1990) The potential of the pitfalls of molecular genetics for psychiatry: an overview. In: Bucyzhenkov V, Christen Y, Prilipko L (eds) Genetic approaches in the prevention of mental disorders. Springer, Berlin Heidelberg New York, pp 107–115
Murray TH (1990) Human gene therapy, the public, and public policy. Hum Gene Ther 1: 49–54
Nelson DL, Chang SMV et al. (1986) Gene replacement therapy for inborn errors of purine metabolism. Cold Spring Harbor Symp Quant Biol 51: 1065–1071
Orkin SH, Williams DA (1988) Gene therapy of somatic cells: status and prospects. Progr Med Genet [New Ser] 7: 130–142
Parkman R (1986) The application of bone marrow transplantation to the treatment of genetic diseases. Science 232: 1373–1378

Phillips JA III (1988) Clinical applications of gene mapping and diagnosis. Progr Med Genet [New Ser] 7: 68–99
President's Commission for the Study of Ethical Problems in Medicine and Biomedical and Behavioral Research (1983) Screening and counseling for genetic conditions. US Government Printing Office, Washington
Rauskolb R (1979) Fetoscopy in Western Europe. A short preliminary survey. In: Murken JD, Stengel-Rutkowski S, Schwinger E (eds) Prenatal diagnosis. Enke, Stuttgart, pp 193–201
Rifkin J (1983) Algeny. Viking, New York
Riordan JR, Rommens JM, Kerem B-S, et al. (1989) Identification of the cystic fibrosis gene: cloning and characterization of complementary DNA. Science 245: 1066–1073
Roberts L (1990) To test or not to test? Science 247: 17–18
Rommens JM, Iannuzzi MC, Kerem B-S, et al. (1989) Identification of the cystic fibrosis gene: chromosome walking and jumping. Science 245: 1059–1065
Schmidtke J (1986) Diagnosis of genetic disease using recombinant DNA. Hum Genet 73: 1–11
Scrimgeour JB (1973) Amniocentesis: technique and complications. In: Emery AEH (ed) Antenatal diagnosis of genetic disease. Livingstone, Edinburgh, pp 11–39
Scrimgeour JB (1979) Fetoscopy. The future. In: Murken JD, Stengel-Rutkowski S, Schwinger E (eds) Prenatal diagnosis. Enke, Stuttgart, pp 202–205
Selden RF, Skoskiewicz MJ, et al. (1987) Implantation of genetically engineered fibroblasts into mice: implications for gene therapy. Science 236: 714–718
Simoni G, Brambati B, Danesino G, et al. (1984) Diagnostic application of first trimester trophoblast sampling in 100 pregnancies. Hum Genet 66: 252–259
Sonneborn TM (ed) (1965) The control of human heredity and evolution. McMillan, New York
Sorge J, Kuhl W, West C, Beutler E (1986) Gaucher disease: retrovirus-mediated correction of the enzymatic defect in cultured cells. Cold Spring Harbor Symp Quant Biol 51: 1041–1046
Stanbury JB, Wyngaarden JB, Fredrickson DS (1983) The Metabolic bases of inherited disease, 4th edn. McGraw-Hill, New York
Super M, Schwarz MJ, Sardhavalla IB (1990) CF screening. Nature 344: 113–114
Thomas KR, Capecchi MR (1986) Targeting of genes to specific sites in the mammalian genome. Cold Spring Harbor Symp Quant Biol 51: 1101–1113
Varmus HE (1984) The molecular genetics of cellular oncogenes. Annu Rev Genet 18: 553–612
Ward RHT, Modell B, Petrou M, Karagozlu F, Douratsos E (1983) Method for sampling chorionic villi in first trimester of pregnancy under guidance of real time ultrasound. Br Med J 286: 1542–1544
Wertz DC (1990) Ethical issues in the application of knowledge from molecular genetics to mental disorders. In: Bulyzhenkov V, Christen Y, Prilipko L (eds) Genetic approaches in the prevention of mental disorders. Springer, Berlin Heidelberg New York, pp 92–106
White R, Lalouel J-M (1988) Sets of linked genetic markers for human chromosomes. Annu Rev Genet 22: 259–279
Willis RC, Jolly DJ, et al. (1984) Partial phenotypic correction of human Lesch-Nyan (hypoxanthine-guanine-phosphoribosyl-transferase-deficient) lymphoblasts with a transmissible retroviral vector. J Biol Chem 259: 7842–7849
World Council of Churches (1982) Manipulating life: ethical issues in genetic engineering. Church and Society, World Council of Churches, Geneva
World Health Organization (1972) Genetic disorders: prevention, treatment and rehabilitation. WHO Tech Rep Ser 497

Molecular Genetic Research in Psychiatry: Ethical Problems

V. I. Trubnikov

The two main ethical problems in molecular genetic studies of mental disorders are:
1. the development of ethical and legal principles and
2. control over their proper use. To solve these problems, one should proceed from the free and unique nature of personality, guaranteed by moral and legal norms of the society.

Molecular genetic research should be limited to the extent of being in line with legal and ethical norms and should be a result of free and unrestrained decision-making. The inclusion and exclusion criteria should be observed from the preliminary stages of the study. Ethical guidance should be supplemented by legal control; it should be clearly defined so as to establish a correspondence between the aims of molecular genetic research and the interests and rights of family members and those of the society. It will make ethical norms absolutely indispensable, and not a matter of "good manners."

The specificity of ethics in molecular genetic research in psychiatry is predetermined by the following:
1. the specific nature of the examination, diagnosis and treatment. The specificity examination is due to the fact that mental disorders are the subject of investigation and treatment. Thus, personal contacts between the doctor and the patient, as well as an atmosphere of confidence and cooperation, are basic guarantees of success. A broad spectrum of the concerns of the patients and their relatives should be taken into consideration. One should look into how the illness affects the patient's social and professional status and his family relationships, what can serve as a stimulus for examination and treatment and what can cause irritation.
2. The peculiarities of the patients' attitude toward their illness and examination.
   a) The reaction of the patients and their relatives corresponds to their condition and the information offered.
   b) The patients and their relatives ignore the illness and the investigation and sound optimistic about the prognosis.
   c) The patients show a negative attitude towards the investigation and refuse to see a doctor.
   d) The patients and their relatives sound positive about the investigation due to an opportunity to see different doctors. Some patients see the

Srám, Bulyzhenkov, Prilipko, Christen (Eds.)
Ethical Issues of Molecular Genetics in Psychiatry

investigation as a way to exclude the presence of an illness; some seek sympathy, attention or better care, while others, especially those who overestimate the severity of their condition, look for confirmation of their own judgments.

3. The patients' social status and their moral and psychological environment. Present-day psychopharmacology enables many patients to avoid a long stay in a psychiatric hospital and to receive maintenance therapies through out-patient service. As a result, a number of ethical problems arise in "doctor-patient-family-society" relationships. Mental illness often causes conflict situations due to an inappropriate attitude of people toward mental patients. Psychiatrists should help the patients and their families to get a clear idea about the mental disorder and about participation in molecular genetic studies, thereby facilitating diagnosis, treatment and social rehabilitation processes, as well as their participation in the investigation. The patients' mental state can fluctuate and their behavior can lead to unpredictable situations, so a researcher should adjust quickly to a new situation.
4. The specific character of confidentiality. Keeping in mind the patient's moral and psychological environment, it is imperative to guarantee full confidentiality with regard to both the patients and their families.
5. The fact that the patients and their relatives get the impression that they are not being treated, but are only involved in research.

In developing ethical criteria for molecular and genetic research in psychiatry it seems logical to proceed through the sequence of stages in the investigation. The following format can be used:

1. the search for informative families;
2. examination of the families;
3. collection of DNA samples and immortalized cell lines;
4. molecular genetic analysis of DNA samples;
5. predisposition gene mapping;
6. discovering predisposition gene carriers using molecular methods; and
7. the application of the information obtained during the study for medical genetic counseling of families with mental pathology. Each stage requires the observance of its own ethical rules.

Analysed here are the ethical aspects of the search and examination of informative families for the analysis of gene linkage and the isolation of DNA samples. These ethical problems can be qualified as those of the "current" study. They appear as a result of those actions which take place in the course of the study. The first group of problems includes contact with the family. Getting the informed consent of the patient is, to a great extent, an ideal situation. Therefore, it is not sufficient to have one interview; there should be close contact between the doctor and the patient and his family. While communicating with the family, the researcher should take into account personality traits, social interests, psychological motivation and the interviewed person's attitude toward the investigation. The following points seem to be important:

1. The doctor's appearance and behavior are essential for emotional contact, mutual confidence and respect.

2. The doctor's explanations should be convincing and easy to understand.
3. His arguments should be vivid and logical.
4. The doctor must show a friendly attitude toward the patients and their families.
5. The doctor must be sure that the participation of the family in molecular genetic investigation is useful for them.
6. The safety of the investigation should be made clear.

The advantages to the family of participation in the investigation should be described: they get the privilege of being consulted by top professionals being admitted to the best hospitals and getting paid on the same terms as blood donors.

Problems in collecting family data for molecular genetic research, for example in schizophrenia (Moscow) have been due to the following:
1. Around 10% of schizophrenic families have first-degree relatives affected.
2. Large families make up 10% of these families.
3. Both parents are not always available for examination in large families (they are dead or divorced or live in different cities). Complete families make up around 50% of the total.
4. There is a high rate of refusal among the complete large families as a result of a mental disorder.

The consent rate is around 20%. Thus, in large cities, only one or two families are examined, after 1000–5000 records have been studied and 20 or 30 families have been interviewed. The examination of family members comprises a complex of measures. Top professional clinicians, psychologists, neurophysiologists and geneticists are invited, and computerized tomography is conducted. These data are indispensable both for diagnosis and further studies dealing with genetic linkage analysis, because there is a higher probability of detecting linkage when the phenotype is viewed through elementary traits. Disposable instruments are used for blood collecting, with further extraction of DNA and lymphoblastoid cell lines.

Thus, such investigations (except blood collecting) have a different degree of significance for the diagnosis. They are absolutely indispensable in manifest cases and possibly important in unaffected relatives. It is absolutely necessary to guarantee:
1. freedom of choice (whether to participate in the investigation or not);
2. safety of investigation;
3. the absence of deteriorating influences on the treatment.

At the same time, compensation is determined by the existing legal regulations and depends on the researchers' resources and possibilities. The important ethical element of the investigation is ethical stability. The unity of the laboratory and clinic (finances, staff) ensures the ethical stability of research. Investigations carried out in conjunction with other institutions require additional ethical controls. Appropriate control is ensured by local ethical committees set up at psychiatric departments of the medical institutes involved.

Ethical Issues Arising in Molecular Genetics in Developing Countries

I. C. Verma

Genetic disorders in developing countries

Let us consider the current status of genetics in developing countries, taking India as an example. Admittedly, genetic disorders are not, at present, significant causes of morbidity and mortality in India or in other developing countries. This is in sharp contrast to the situation in the developed countries. The difference in the pattern of disease is reflected in the difference between their population structures (Scriver 1976). The demographic profile in a developed country resembles a cylinder, while in a developing country it is like a pyramid (Fig. 1). In developing countries, the large number of births give rise to a broad base, while the tip is formed by very few people in the older age groups. The pyramidal shape arises from mortality in each age group resulting from infectious and parasitic disorders (Verma 1986a). The cylindrical shape in a developed country is due to a very low mortality in childhood, resulting in the predominance of genetic, degenerative, and malignant disorders in older age groups.

Dr. Modell (1985) suggested that genetic disorders become important once the infant mortality rate falls to less than 50 per 1 000 and the birth rate to less than 20 per 1 000. This is shown graphically in Figure 2, which correlates the

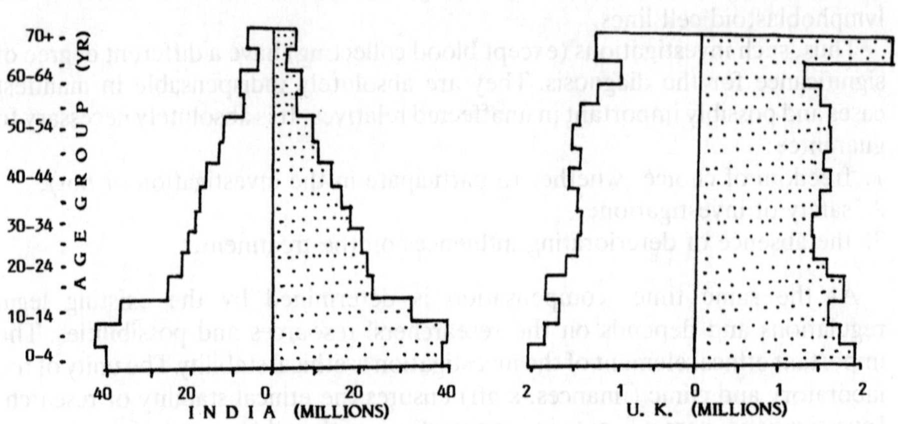

Fig. 1. Profiles of population structures of India and the United Kingdom (UK)

Srám, Bulyzhenkov, Prilipko, Christen (Eds.)
Ethical Issues of Molecular Genetics in Psychiatry

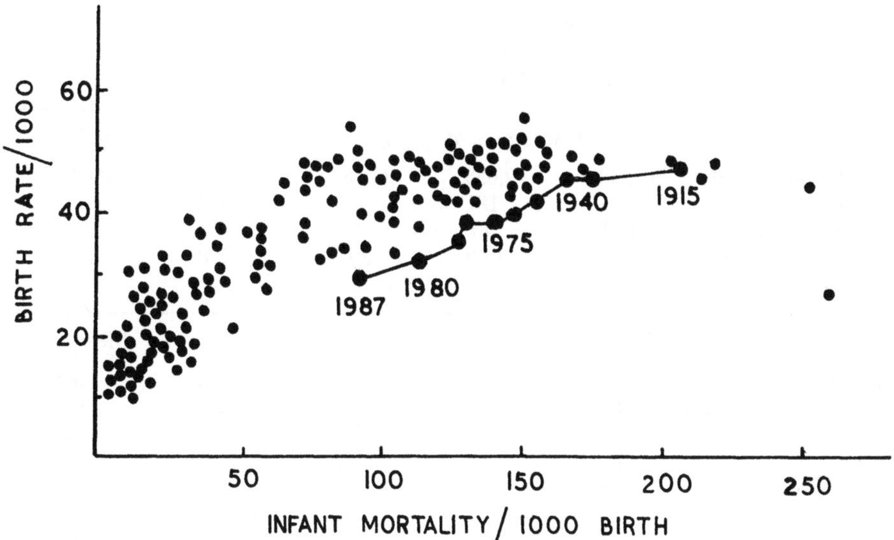

Fig. 2. The infant mortality rate plotted against birth rate. The line traces the infant mortality rate for India in different years. The developed countries fall in the left lower corner, while the developing countries are represented in the right upper corner.

infant mortality rate to birth rate in different countries. The infant mortality rate in India currently stands at 85 per 1 000. However, the experience in India shows that genetic disorders are a significant problem, in spite of a high frequency of communicable diseases. For a variety of reasons, such as high prevalence of consanguinity, falciparum malaria in the distant past, high birth rate and a large population, and lack of rehabilitative facilities, the burden of genetic disorders is even greater in India, as compared with the developed countries (Verma 1986b). Therefore, genetic disorders deserve attention, even in developing countries, to reduce the burden that they place on the patients and their families.

The genetic disorders which are presently of importance in India are as follows:

1. congenital malformations;
2. chromosomal disorders, especially Down syndrome;
3. genetic causes of mental retardation, for example fragile X syndrome and inborn errors of metabolism;
4. hereditary anemias;
5. muscular dystrophy;
6. neurologic and other psychiatric disorders.

Psychiatric Disorders and Their Care

A multicentre study was conducted by the Indian Council of Medical Research in four centres, using well-standardized proformas; it showed the total prevalence of severe psychiatric morbidity in rural areas to be 9.5 per 1 000 (Table 1; Isaac 1988). The mean prevalence of epilepsy was 3.5, manic depressive psychosis 2.9,

Table 1. Prevalence of severe psychiatric morbidity in rural areas in India

	Bangalore	Baroda	Calcutta	Patiala	
	Characteristics of study group				Total
Population studied	35,548	39,655	34,582	39,595	146,380
No. of patients	395	181	287	517	1,380
Rate per 1000	11.1	4.6	8.3	14.1	9.8
% patients managed by the PHC[a]	18.2	19.9	20.2	12.8	16.8
	Disorder[b]				Mean
Epilepsy	7.8	1.3	1.7	3.2	3.5
Organic brain damage	0.1	0.6	0.6	2.4	0.9
Schizophrenia	1.8	1.8	2.1	3.1	2.2
Mania	0.6	0.4	0.2	1.4	0.6
Depressive psychosis	0.8	0.6	3.7	4.1	2.3
All disorders	11.1	4.6	8.3	14.1	9.5

[a] PHC, primary health centre
[b] Prevalence per 1000

schizophrenia 2.2 and organic brain damage 0.9 per 1000. Other community studies of various psychiatric disorders found a range of 18 to 370 per 1000, depending upon the methodology used and thoroughness of the study (National Mental Health Programme for India 1989). The mean prevalence in 10 studies was estimated to be 90.3 per 1000 (Fig. 3). In the hospital clinics (Sethi and Manchanda; Verma et al. 1980), schizophrenia is the commonest cause of

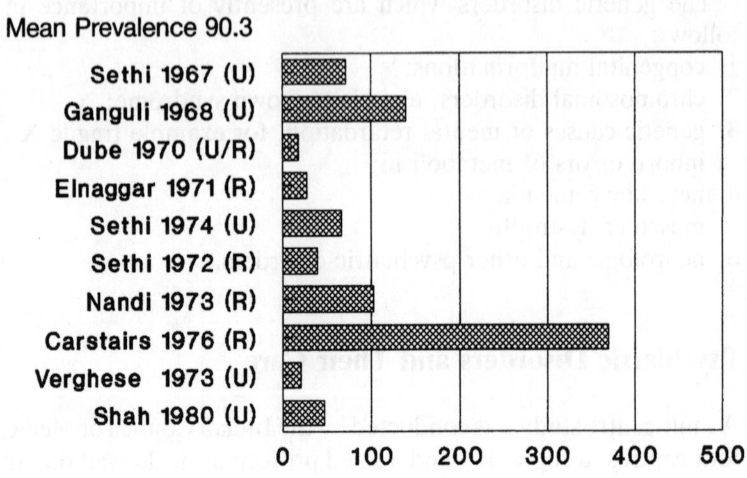

Fig. 3. Prevalence of psychiatric disease in community studies in India. Data are modified from the National Mental Health Programme for India (1989). U, urban; R, rural.

referral (about 50%), followed by manic depressive psychosis (30%), organic psychosis (7%) and neurosis (6%).

It is apparent that psychiatric morbidity is high and, due to the large population in India, the total number of psychiatric patients is more than the total population of many European countries. For example, using the data given above, the number of persons who have severe psychiatric morbidity is estimated to be 700,000; those with all types of psychiatric disorders number 7.2 million.

Realising the unequal distribution of health services in India, with its greater urban bias, the Government of India has opted for a primary health care approach to serve the masses of India (National Mental Health Programme for India 1989). In the field of psychiatric disorders, community health centres have been established in a large number of places, depicted in Figure 4.

Facilities for Molecular Genetic Studies

Molecular genetic techniques have not been employed much for the management of psychiatric patients in India, firstly because the location of genes which are responsible for psychiatric disorders is not precisely known, although knowledge in this regard is increasing, and secondly because there are only a few molecular genetic laboratories in India, mostly located in the universities and engaged in basic research. Only recently have molecular genetic laboratories been established in the hospitals to provide service to patients with genetic disease.

The present trend toward simplifying the molecular genetic techniques – for example, by PCR technology and non-isotopic probes – will hopefully lead to the application of molecular genetic techniques to manage patients with psychiatric disorders in the near future in developing countries like India. The disorders for which molecular genetic techniques would have the greatest utility in India and other developing countries are listed in Table 2.

Factors Influencing Ethical Decisions

The Patient

Examining the factors which influence decisions regarding ethical issues in developing countries, it should be clear that the key figure in this process is the patient, rather than the physician. The patient makes ethical decisions after interacting with the physician. The actions of the patient are moulded by pressures from family members, such as parents and grandparents. This is especially so in developing countries because of the joint or extended family system. Close relatives often exert so much moral pressure that ordinary couples find it difficult to resist.

Fig. 4. Map of India showing the location of community mental health centres.

Religion

In developing countries religion continues to have a major influence on ethical decisions, especially in the rural communities. However, the Hindu religion is not dogmatic, unlike the Muslim or the Christian religion. It emphasizes self-realization, thus leaving the choice of action in a particular instance to the individual, although it is expected that the higher morality is kept in mind while selecting a course of action. It is not surprising, therefore, if the patient opts to reduce the economic burden that might be imposed by the genetic disorder by

Table 2. Application of molecular genetics in developing countries

A. *Early diagnosis and management*
1. Neurologic disorders
   - Neurofibromatosis
   - Hereditary ataxias
   - Huntington's chorea
   - Myotonic dystrophy
2. Psychiatric disorders
   - Schizophrenia
   - Affective disorders
   - Alzheimer's disease
   - Fragile X syndrome

B. *Prenatal diagnosis*
1. Hereditary anemias:
   Beta thalassemia, Sickle cell anemia
2. Duchenne muscular dystrophy
3. Fragile X syndrome
4. Hemophilias A and B
5. Determination of sex: X-linked disease, ? Social reasons

choosing abortion of an affected fetus, or by not favouring prolonging the life of an infant who is likely to be severely handicapped.

Socio-economic Burden of Disease

In view of the widespread poverty, another factor which weighs heavily in the choice of the patient is the socio-economic burden of the disease in question. This factor often overrides religious and cultural influences, especially for those exposed to Western thought.

The Law

The patient and the physician are both governed by the law of the land, but its enforcement is often feeble in developing countries. Therefore, laws that are applicable to individuals which only marginally affect the society at large remain unenforced. For example, even if someone in the family or neighbourhood knew about an illegal abortion of a female fetus, it is unlikely that the matter would be reported to the police. Moreover, in many cases where actions are considered unethical, e.g., asking for dowry in the marriage of a son, the girl's parents would not report the matter to the police because it would affect the future of their daughter. Sometimes the law is also anomalous; for example, bigamy is prohibited by law, but the person who lodges a complaint is the first and rightful wife. She may opt not to lodge a complaint because the husband may refuse to give her maintenance, or decline to help her and her children. She could get redress from the courts, but this is a lengthy, drawn-out process which the traditional woman tries to avoid.

Ethical Obligations of Physician

The physician has three ethical obligations:
1. he must understand the basis of his moral choices;
2. he should have a working knowledge of ethical principles; and
3. he must follow a logical plan of ethical decision-making. Unfortunately, the majority of physicians in developing countries lack any formal training in the discipline of bio-ethics. Therefore, these obligations remain unfulfilled.

Physician-Patient Relationship

The medical advice which the physician gives is influenced by the nature of the physician-patient relationship that exists in a particular instance. Most of the government and university hospitals in developing countries like India provide free medical care, so that the relationship is one of service or academics. In certain tribal communities in developing countries the relationship is quasi-religious, and the community doctor performs a social function. However, in most countries, including India, increasingly the model of a business transaction between the patient and the doctor is being followed. This is unfortunate but true. In developing countries this tends to distort the advice rendered by the physician, with the objective of personal monetary gain. For example, in a case of spina bifida, a surgeon in private practice would tend to advise surgery somewhat indiscriminately. Ideally the system of patient care should be based on an internal morality of medicine which imposes obligations on the clinician and expects him to follow the three ethical principles of beneficence, autonomy and justice, keeping in view three important criteria:
1. confidentiality,
2. truth telling, and
3. promise keeping (Pellegrino 1987).

Let us examine each of these principles as they apply in developing countries.

Beneficence

Beneficence requires that the physician act in the best interest of the patient. In genetics we are often concerned with reproduction and future generations. The question thus arises, whose interest should the physician cater to – mother, father, fetus, or the society? For example, in the case of sex selection the parents desire to have a child of their choice. If they then abort a fetus of the unwanted sex, they are serving their interest, but not that of the fetus. In most communities the rights of the fetus are not paramount, but are secondary to those of the mother and father. This is especially true in developing countries.

In developing countries the principle of benificence has to be applied keeping in mind the quality of life as well as the burden which a particular disorder places on the family. For example, in developing countries the burden of rearing a child with spina bifida is very heavy. Most physicians would advise surgery only in very selected cases, such as for those who do not have any disability or have a mild one. They might not recommend surgery if there was paralysis of the bladder and the lower limbs.

Autonomy

This principle respects the patient's moral right to decide for himself. However, very often this principle is in direct conflict with beneficence. The patient may opt for a course which he considers best for him, but this may be different from the perspective of the physician or the society. To what extent unbridled freedom of choice should be allowed is debatable. For example, in the case of sex selection for social reasons, the principle of autonomy would state that the patient has the right to decide to have a child of the desired sex, for example, a boy. Is this ethically wrong or right? In spite of the many laws which have been enacted to protect the rights of women in India (Table 3), discrimination against women continues unabated. The birth of a girl is greeted with silence, and even sorrow. She faces continuous physical and financial neglect in favour of her brothers. It is not surprising, therefore, that women have a low literacy rate and a high morbidity and mortality. The government has clearly decided to prevent the discrimination against women from extending to before birth by regulating the use of amniocentesis and chorionic villus sampling tests, permitting these only for the diagnosis of genetic disease. The government has further banned the use of these techniques for the purpose of abortion of a female fetus. The state of Maharashtra has already introduced a bill for this purpose (Maharashtra Regulation 1988), and the government of India is planning to pass a national bill on the same issue.

The autonomy of the individual is operative in a real sense only when the person is sufficiently knowledgeable to take appropriate decisions. How far should one follow this principle in developing countries where literacy rates are low? Under the aegis of the Indian Council of Medical Research a five-centre study was carried out in Delhi, Bombay, Lucknow, Pune and Bangalore to examine the knowledge and attitudes of the patients who received genetic counselling (Indian Council of Medical Research 1989). It was found that most of the patients had no idea about the cause of the disorder in their offspring (40 %), while many ascribed it to God's will (33 %) or their bad luck (Fig. 5). Only 30 % of those who attended the clinics knew about genetics and heredity, while 47 % understood it partially (Fig. 6). Therefore, in such a situation directive counselling can be permitted to a certain extent. In counselling parents who have no or poor knowledge of genetics, it is the practice in our genetic unit to state what other parents in their situation would do rather than giving straightforward directive counselling.

Table 3. The Indian constitution in support of the rights of women

Article 14: The state shall not deny to any person equality before the law, or the equal protection of the laws within the territory of India.

Article 15: The state shall not discriminate against any citizen on grounds only of religion, race, caste, sex, place of birth or any of them.

Article 16: No citizen shall, on grounds only of religion, race, caste, sex, descent, place of birth, residence or any of them, be ineligible for, or discriminated against in respect of, any employment or office under the State.

The Dowry Prohibition Act 1961, Rules, 1985, Amendment 1986. If any person gives or takes or abets the giving or taking of dowry, he shall be punishable with imprisonment for a term which will not be less than 5 years, and with fine not less than Rs. 15,000/– or the amount of value of such dowry.

* If any person demands, directly or indirectly, from the parents or other relatives or guardian of a bride any dowry he shall be punishable with imprisonment for not less than 6 months, but which may extend to two years, with fine which may extend to Rs. 10,000.
* Every offence under this Act shall be nonbailable and noncompoundable.
* Where any person is prosecuted under this Act the burden of proving that he had not committed an offence shall be on him.

Amendment 1986: where the death of a woman is caused by any bodily injury within 7 years of her marriage, and it is shown that she was subjected to cruelty or harassment by her husband or any relatives such death shall be called Dowry death, punishable with imprisonment of not less than seven years, extendable to life.

Under the Hindu Succession Act 1956, the property of a male or female Hindu dying intestate, shall devolve firstly upon the heirs in class of the schedule, simultaneously to all of them. Class 1 of the schedule includes son, daughter, widow and mother.

Fortunately, it was observed that, in spite of their initial poor knowledge, genetic counselling brought about a distinct improvement in their understanding of the knowledge of the cause of the disorder (Fig. 5). There was also a remarkable improvement in knowledge of description of the disorder (Fig. 7) and regarding the risk of recurrence after counselling (Fig. 8). Indeed the common reason for not understanding was that the explanations of the doctor were too technical. This highlights the need to develop simple and pictorial methods of counselling to explain the genetic facts to illiterate or poorly literate individuals.

Initially, we were worried that genetic counselling might have a negative effect on marriage in those instances where the female is a carrier of a mutant gene, because of the low status of females in society. However, this fear was unfounded; in 56 % of the cases, counselling had a positive effect. The counselling also helped the couples to cope with the disorder, as 70 % confirmed a beneficial effect, while another 30 % agreed to some effect. It was also observed that the harmful effects of counselling on family life were very low, being observed in only 2.5 % of the patients counselled.

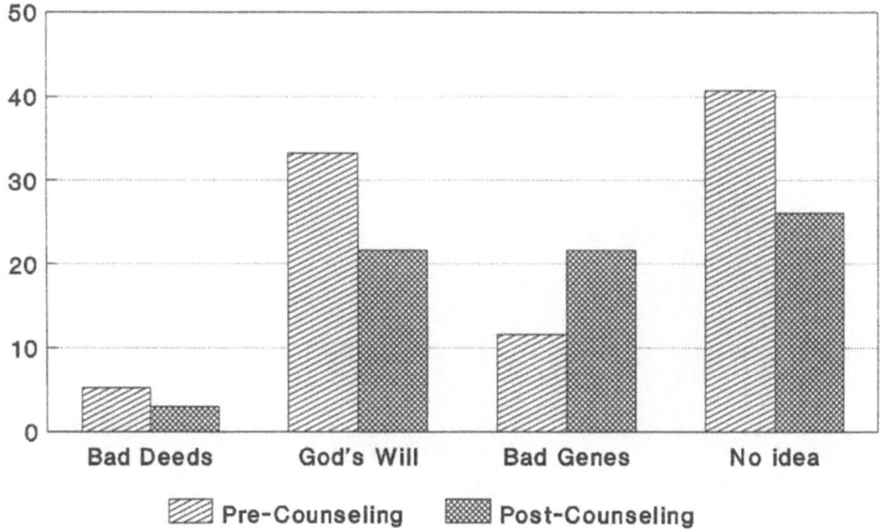

Fig. 5. Knowledge of the genetic counselees in India about the cause of disorder. Data obtained before and after counselling are represented from five centres.

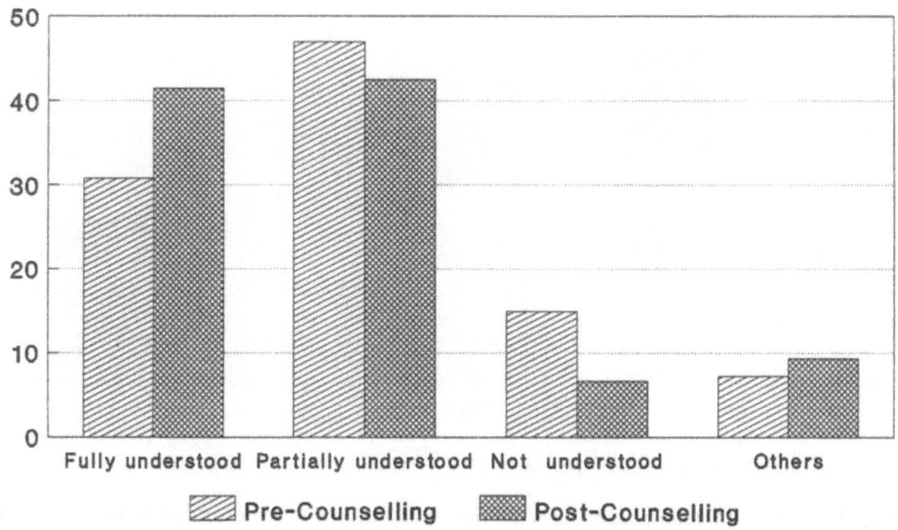

Fig. 6. The counselees' understanding of genetics, both before and after counselling.

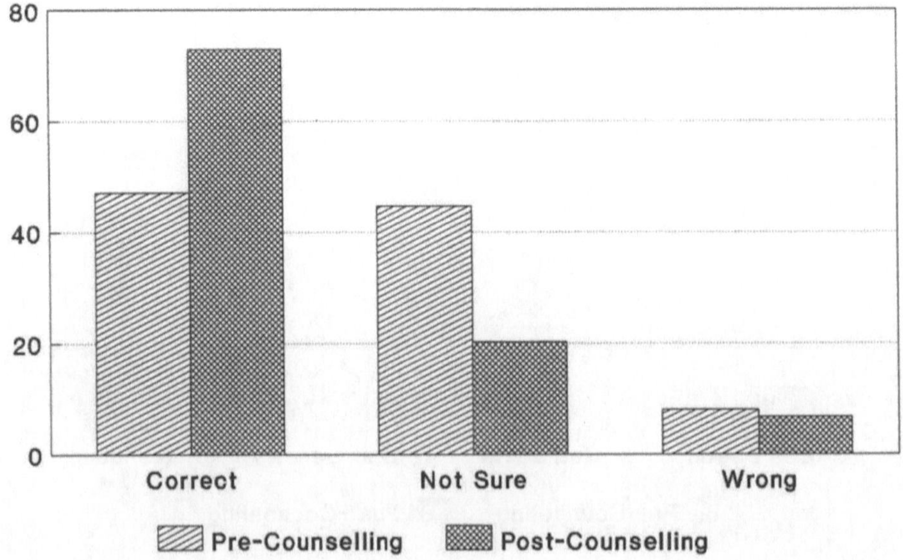

Fig. 7. The counselees' knowledge about the description of the disorder. Note the improvement in correct responses after counselling.

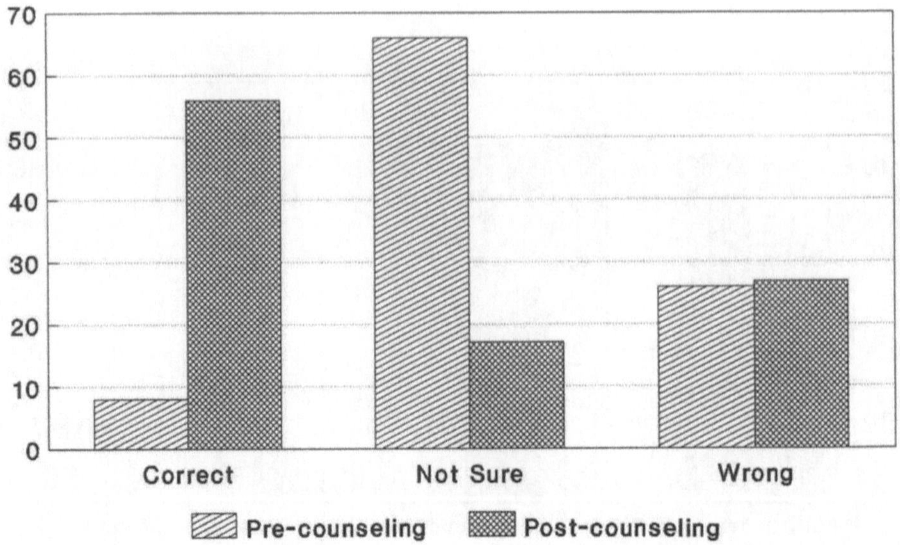

Fig. 8. Counselees' knowledge about the risk of recurrence of the disorder, before and after counselling. Note the remarkable improvement in correct responses, showing the success of genetic counselling.

Justice

This principle is a crucial one because in all developing countries the availability of health services is not equitable, with most of the rural areas being underserved. For example, in the study mentioned above (Indian Council of Medical Research 1989), only 27.7% of the counselees were from the rural areas, although 78% of the people in the general population live in rural areas. This clearly shows that genetic counselling clinics were attended by those in urban or semi-urban areas, and that the vast majority of those living in rural areas were not using these services. The principle of justice, therefore, requires that genetic counselling services be provided through primary health care as near to the homes of the people as possible. Therefore, the establishment of community mental health centres by the Government of India is an appropriate strategy.

Confidentiality and Truthfulness

It is accepted that the physician should be truthful in explaining the facts of the case to the patient, and maintain confidentiality regarding the information gathered about the patient and the family during the course of professional work. However, there are occasions when these principles would need to be relaxed. For example, consider Huntington's disease. This is a neuropsychiatric disorder which usually manifests in the third or fourth decade, after the person has completed his reproduction. Closely linked probes allow presymptomatic diagnosis of this disorder with reasonable accuracy. If one affected person is detected, then the diagnosis could easily be established in other related individuals at an early stage, allowing them to plan their future. But what should the counsellor do with regard to disclosing the information to other relatives if the patient refuses to do so himself or herself? We surveyed the views of genetic counsellors in India on ethics as applied to the practice of medical genetics. Twenty-seven counsellors replied to our questionnaire. In the case of Huntington's disease, 48% of the counsellors favoured disclosing the diagnosis to relatives, even if the client refused to permit such disclosure; 33% favoured disclosing the information even if the relative did not ask for it. Such a situation would also arise if a molecular genetic test becomes available for presymptomatic diagnosis of schizophrenia or manic-depressive psychosis. In India the patient/parents often do not wish to disclose the information that someone in the family is suffering from psychosis. They also tend not to disclose that other family members have similar psychiatric disorders. However, it is the experience of most psychiatrists that, once the physician has gained the confidence of the patient, he or she is willing to disclose the information regarding other family members that are similarly affected.

Another common situation where the issues of confidentiality and truthfulness arises is the disclosure of a parent who is a carrier of a mutant gene/chromosome, be it schizophrenia, neurofibromatosis, tuberous sclerosis or translocation Down syndrome. The problem is stigmatization and the possible repercussions this would have on the marriage. In our study, no consensus emerged on disclosing information regarding which parent is a balanced translocation carrier in Down syndrome.

Ethical Issues in Molecular Genetics

Table 2 depicts the psychiatric, neurologic and genetic disorders for which molecular genetic techniques have the greatest application in developing countries. The ethical issues which arise in the use of these techniques for helping patients with these disorders are briefly discussed here.

Genetic Counselling

The purpose of a genetic consultation is counselling, and that is what the patient is most interested in. Many of the patients who come for genetic counselling in India have very poor knowledge of genetics. Therefore, to assist the patient to make proper decisions regarding reproductive choices, the genetic counsellor would have to explain the inheritance of the disorder, the risk of recurrence and the implications of the molecular genetic tests in simple language to the counselees. More directive counselling is permissible/ethical for illiterate patients. Those who have knowledge of scientific principles can be offered non-directive counselling, as in Western countries.

Prenatal Diagnosis

As explained above, people in developing countries are very concerned with the socio-economic burden of genetic disorders, they readily accept prenatal diagnosis and opt for abortion of the fetus is affected. This is true for those who follow the Hindu religion or other religions derived from it, like the Sikh, the Jain or Buddhist religion. However, traditional Muslims do not accept prenatal diagnosis, although Muslims who are modern (Western) in their outlook do accept it.

In India, as in some other developing and developed countries, there is a great preference to have a male child. A large number of people seek prenatal diagnosis of the sex of fetus for social reasons. Chorionic villus sampling has provided even greater impetus to this practice. What is even more unethical is opting for abortion if the fetus is diagnosed to be female, as discussed above.

Genetic units in government hospitals and institutions are not permitted to carry out prenatal diagnosis of sex for social reasons. It is the practice in these institutions, while reporting results of prenatal diagnostic tests, not to disclose the sex of the fetus but only to state that the fetus is normal/abnormal. The majority of patients readily accept this, and do not insist on knowing the sex of the fetus.

Therefore, the government of India is planning to introduce a bill to regulate prenatal diagnostic tests so that they are performed only for diagnosis of genetic disease and to prevent their misuse for the purpose of fetal sexing, leading to abortion of a female fetus. The state of Maharashtra has already introduced a bill to this effect (Maharashtra Regulation 1988). It is recognized that legislation by itself cannot solve deep-rooted social problems, however, such legal sanctions are essential in developing countries to provide the push and the educative factor to mould public opinion.

One great advantage of molecular genetic techniques has been their application for prenatal diagnosis of disorders which were hitherto not diagnosable prenatally, such as Duchenne muscular dystrophy, or were difficult to diagnose, such as beta thalassemia and hemophilia A and B. These tests have been readily accepted, and no ethical problems have ensued. The only remaining problem is their cost; some of the samples for these tests have to be sent abroad, the cost of Rs. 10,000 – Rs. 15,000 (US $60–90) is prohibitively expensive. There is a need to develop regional centres for molecular genetic services in India. The above comments would also apply to molecular genetic tests for psychiatric disorders, when these become feasible.

Genetic screening

Genetic screening can involve screening of the general population (mass screening) or be restricted to a high risk group. Mass screening is commonly confined to a particular age group, e.g., newborns or pregnant women, etc. In most developing countries, the level of biotechnology and economic considerations do not permit genetic screening programs involving whole populations. Presently this may be justified only for hypothyroidism in the newborn period, in areas endemic for iodine deficiency goitre, because studies have shown that 1.5–13.3 % of the newborns in such areas suffer from chemical hypothyroidism. Many of these children exhibit a whole range of neuro-developmental disorders later on (Kochupillai and Pandav 1987). A case could be made for newborn screening for sporadic hypothyroidism, although economic considerations have prevented initiation of such programs in developing countries. Although the frequency of amino acid disorders leading to mental retardation is similar to that in the West (Appaji et al. 1990), the treatment with special diet is currently prohibitively expensive for the vast majority of people. Therefore, amino acid screening is at present not feasible in developing countries.

The only other disorder where population screening may be justified in developing countries is hereditary anemia. Of course, the age group which should be screened for carrier status has to be carefully selected. A case could be made for screening for sickle cell disease in the newborn period, because if a homozygote is detected he/she can be provided with penicillin prophylaxis or immunization against pneumococcal infection. This has been shown to prevent early death. However, screening for beta thalassaemia in developing countries should be delayed until after marriage. The reason is that the large majority of marriages in these countries are arranged by the parents, and carrier screening programs before marriage would tend to stigmatize the individual. Therefore, delaying carrier testing until after marriage would remove the fear of stigmatization. The test should preferably be done before pregnancy. Alternatively, the mother could be tested early in pregnancy and, if she proved to be a carrier, the husband could be tested. In the event that both husband and wife were carriers, prenatal diagnosis could be offered on chorionic villi samples obtained at 9–10 weeks of pregnancy using molecular genetic techniques. Similar principles would apply to genetic screening for psychiatric disorders, once the responsible genes are identified and their chromosomal locations are determined.

The facts mentioned above highlight the need for starting genetic screening only after proper ethical considerations, after appropriate information has been disseminated to the people, and after arrangements have been made to provide them continued support.

Conclusions

In the ultimate analysis, ethical judgement involves not a choice between an absolute right or wrong, but a conflict between competing "rights." This is what democracy is all about, and the choice must remain with the people who are affected.

References

Appaji Rao N, Rama Devi AR, Savithri HS, Rao SV, Bittles AH (1988) Neonatal screening for aminoacidemias in Karnataka, South India. Clin Genet 34: 60–63

Indian Council of Medical Research (1989) ICMR Task Force on Human Genetics (1985–1987): report on "genetic counselling." Indian Council of Medical Research

Isaac MK (1988) Severe mental morbidity. ICMR Bull 18(12): 126–137

Kochupillai N, Pandav CV (1987) Neonatal chemical hypothyroidism in iodine deficient environments. In: Hetzel BS, Dunn JT, Stanbury JB (eds) Prevention and control of iodine deficiency disorders. Elsevier, Amsterdam, pp 85–93

Maharashtra Regulation (1988) of Use of Prenatal Diagnostic Techniques Act. Government of Maharashtra, Bombay (Maharashtra Act no XV)

Modell B (1985) In: Report of WHO Advisory Committee on hereditary diseases. WHO, Geneva (WHO document AMG/AG/85.10)

National Mental Health Programme for India (1989) Progress report 1982–1988. National Institute of Mental Health and Neurosciences, Bangalore

Pellegrino ED (1987) The anatomy of clinical-ethical judgements in perinatology and neonatology: a substantive and procedural framework. Semin Perinatol 11: 202–209

Scriver C (1976) Genetics, voyage of discovery for every man. Pediatr Res 10: 865–872

Sethi BB, Manchanda RM (1979) Socio-cultural attitudes and psychiatric illness in India. In: Ahuja MMS (ed) Progress in clinical medicine, 3rd series. Heinemann, Delhi, pp 532–550

Verma IC (1986a) Medical genetics in India. Indian J Pediatr 53: 437–440

Verma IC (1986b) Genetic disorders need more attention in developing countries. World Health Forum 6: 69–70

Verma VK, Wig NN, Srinivasa Murthy R, Misra AK (1980) Socio-demographic correlates of schizophrenia, affective psychosis and neurosis in a clinic in India. Acta Psychiatr Scand 61: 404–412

Religious Approaches to the Results of Genetic Research

C. de Filippi

Introductory Notes

While I thank the World Health Organization (WHO) and the IPSEN Foundation for having invited me to this meeting in the heart of a Europe which is becoming more and more united, I feel the need to define more clearly the quality of my contribution to avoid possible misunderstandings. Although I have completed my philosophical and theological studies as a Roman-Catholic priest and gained a degree in both, I have no official charge, nor do I represent any institution. The following thoughts derive from my personal interest in burning questions of the present day and they are expressions of personal convictions, which I hope are still developing in my own mind.

I hold it necessary to clarify at once that a *religious approach* is not to be identified nor is it synonymous with an *ethical approach*. The latter refers directly to a code of norms, while the former refers to a consciousness of a relationship of the human person and of the world to a transcendent Being.

Main Types of Religiousness

From the very rich and abundant library dealing with studies and inquiries about religious phenomenon during the spaces and times of mankind I think that we can define – given the risk of simplifying matters too much – two major types of religiousness, i.e., expressions in words and acts of human religious sentiment. I will define them as naturalistic and personalistic religions.

From the abundant stories handed down by many peoples, great and small, throughout the world, we can gather an account which presents the living world in deep unity with all of its component parts. These accounts or myths, whose principal figure is normally a woman, are the spiritual expression of an agrarian civilization. Woman, figure and symbol of Mother Earth, is the subject of these accounts, which are mankind's answer to the question of the meaning of the rhythm between fecundity and decadence, between life and death. In the lunar and menstrual cycles and in the succession of seasons and generations mankind recognizes a universal law in which he himself is involved and to which he tries, by means of myths and rites, not only to give a meaning but also to assume it consciously. The myth is the expression of this vision of the world (Weltan-

Srám, Bulyzhenkov, Prilipko, Christen (Eds.)
Ethical Issues of Molecular Genetics in Psychiatry

schauung) that we could call "cosmobiology." It is not only the human interpretation of natural events, the most important of which are birth and death, but is even the acknowledgement of the universal and absolute value of the life-death rhythm and the will to participate fully and consciously in it.

In the battle between life and death men have to give their own contribution to the victory of life in this very moment of the circular flow of history.

But also in this framework of the naturalistic myth there is a symbol which opens to a further understanding, the Virgin Mother. In this symbol mankind recognizes that the fertility of the earth precedes the intervention of man, his activity, and his technology (agriculture and breeding). If the myth is also an account of a fundamental human experience, it recognizes that mankind is not the aim of the world, but the world is a complete entity to whose continuity of life man is called to give his contribution.

From the same treasure of tales and myths we can elucidate a second complex of accounts which gives a different interpretation of man and of the world based on human religious experience. This complex of accounts is not so abundant and universal as the naturalistic myths, but it is no less essential to a correct description of human religious experience. These are myths and tales which develop around two basic symbolisms, the ethical and the cosmic.

The first develops in the consciousness of a supreme Being, which is before the human person as a personal and authoritative presence: he is Father and Lord of man and of all living beings. He sees all, he hears all, he knows all. Before him man has to be responsible for his own actions, the good and the bad. He is a judge; he rewards and he punishes. Lord of the animals, creator of the world, he is not a prescientific explanation of the existing world, but he reflects the human consciousness that man does not own the world, but must use it moderately, according to his needs, because the world belongs to another. Creation myths have precisely this intention: as man is not creator so he is not lord of the creation. Within the order of this world, which is *cosmos* and not *chaos*, man faces a Being who transcends this order because he has posited it.

The cosmic symbolism suggests for this world a structure according to a heavenly, invisible order. This cosmic order takes its shape along an ascending axis: the sacred mountain and the sanctuary on it are the centres around which this world is shaped – no more chaos, disorder and darkness but cosmos, order and light. And every place where man meets the transcendent Being becomes a new "ombelicus mundi," a new point of orientation and of ordering.

Evidently these two types of religiousness refer not only to two different phases of human history – hunting/gathering and agriculture/breeding – they refer also to two different conceptions of man. The naturalistic religions invite man to participate in the great and mysterious dance of life and death, to participate in the circular rhythm of their battle, so that life may always triumph over death.

Personalistic religions invite man to be himself, to become aware of the good but severe presence of a Lord, to become aware of living in complete trust and in necessary submission to an Other, who is different from the world in which man lives. It is evident that the three great monotheistic religions of the world, Judaism, Christianity and Islam, belong to this category of religiousness.

I think that we should add to the previous types of religiousness a third one, philosophical religion, although it is difficult to define. While myth is a first step – but not a naive one – to shape the sense of man's life in this world, philosophy is a second effort to give order and meaning to this world. They are two different languages, whose difference probably derives from a greater or lesser distance from immediate experience. I would call the philosophical thought a double meditation. But what I want to stress is that even philosophy often works with the supposition of a god. Of course: there is a great difference between the God of Abraham, Isaac and Jacob and the God of philosophers, as Pascal says. Man meets the first one in his concrete experience of life, in his personal history or in the history of his people. The second one seems to be the result of a reflecting thought, the ripe fruit of an ontological or metaphysical research. This type of god seems to be born by requirement of human reason, as we see in the philosophy of Aristotle, not by personal experience.

After this very quick and therefore very general view of religious phenomenon we may ask, what does the word religion signify? We could answer that, with the word religion, we point out that human attitude which expresses itself in words and acts and which grows in man's heart as the consciousness that he himself is not the ultimate giver of life and of meaning in this world. Man experiences that there is somebody else beyond this world of our senses to whom he owes responsibility. Or, on the other hand, man experiences this world of our senses as a living unity and he has to live in harmony with this unity.

God is the Other or the All

Usually the problem of *nature* belongs more to philosophical reflection than to religious thought. Nevertheless we can also put the question within a religious framework. What is the nature of a living being? Answer: to reach its aim through suitable means. But this is an answer which produces new questions about the aim and the means. If we set the question in the context of naturalistic religions the answer could be, for instance, that the nature of man is to participate in the circular flow of life and death with all those acts which contribute to the life of the whole living unity. In the context of personalistic religions the answer could be that the nature of man is to obey the Lord's voice, to shape continually the *cosmos* in opposition to the *chaos*.

By doing this we see that the word nature assumes a double meaning,
1. the nature of a single living being, and a more general one
2. that nature is the world not yet changed by man.

This latter meaning is the one used by the "green" political organisations. But my intention is to stress two possible meanings of this word in religious language which are important for the main point of my report.
1. Fixed nature: World and man have a fixed structure which is given from the beginning. This fixed order or nature has to be maintained and strengthened. The idea of change or even of an evolution is excluded.

2. Created nature: The structure of man and of the world depends on the Creator's will. It has its foundation in God's will but it is not identical with nature as it de facto is. There is a difference which the religious man is called to fulfill or to eliminate. I am just thinking of the prophetic accusation against the rich who are rich at the expense of the poor. I am thinking of the messianio promises which announce a mankind freed from diseases. In this vision the nature of man and of the world is not something behind us but something in front of us, something to be built.

The Axes of Religious Experience

I do not wish to develop this topic here, but I am convinced that it is methodologically correct to distinguish three possible levels in the complexity of religious phenomenon. And this is a very important matter, especially when two or more spheres of human experience and activity are compared, as for instance religion and genetics.

First of all we have the testimony of religious practice during the course of the history of mankind up to today. On this level religion is often lived as a traditional heritage, not or not yet personally assumed. It is practice, not conviction.

Religious experience is very often reflected on. Very often, it is shaped in a comprehensive system of fundamental affirmations, meanings and values. As a non-religious world conception can be called ideology, in the same way each religion has or is a theological structure, more or less flexible.

Within a theological structure there is normally a limited area which deals with *norms of behaviour*. Ethics and morals have as the specific object of their interest human behaviour related to a code of recognized values and norms. Even more than theological structures, ethical or moral structures can change in the course of history. Effectively the ethical or moral reflection is stimulated by or stimulates modern results of science on account of their effects on life, human or not human.

Genetics as Knowledge and Action

It may seem strange to ask the question, "What does genetics signify?" during a meeting like this, but I hope you agree with me that it is opportune to describe as precisely as possible a field of human activity when we wish to compare it with other fields. I hope to be not too wrong in defining genetics as that science which tries to discover how the transmission of hereditary factors happens within a population. In an eminent way genetics is "bio-logia," the "science of life," just as there are sciences of the earth and sciences of man.

As long as genetics researches with proper means how a phenomenon happens, i.e., the transmission of hereditary factors, it has to benefit from the same freedom as the human spirit, which always wishes to know more and better the laws inscribed in this world. Theoretically we could say that there are no limits to scientific research. Nevertheless, in each science, and in genetics too, pure scientific research is not a neutral action, outside the sphere of what we call the ethical imperative. It has been many years since we have lost the naive vision of a pure science. Scientific research is mostly conditioned by politics and economics. But even freed from these external conditionings, scientific research is directed to practical applications.

Each of us knows from personal and historical experience that knowledge is not only an aim in itself, it is not only satisfaction of the spirit. It is deeply linked with knowing how to act, with "being able to do," with capacity, with power. He who knows can act. He who knows can act more than he who does not know.

When scientific research passes from an hypothetically pure level of research to an operative level, it confronts the ethical question. It is the same ethical question which summons each human action: that freedom which the human spirit may ask for itself is a freedom directed to the goodness of knowledge. But the same freedom has to be limited when its subject, in this case the researching scientist, begins to work on the reality of the world and of man. No scientist should avoid the possible questions: Why do you do this? How do you do this? Why do you wish to change the nature of this world?

Religious Attitude As a Demand for Ethical Responsibility

Remembering what has been said about the historical types of religiousness, but also looking at the religious phenomenon in our days, we see a plurality of religious behaviour. Each of them has its own conception of the world. Constituent elements of the religious conceptions are the relationships between man and god, between man and the world, and between man and man. The quality of approach and the evaluation of the results of each science, even of genetics, depend on the value which is given to these relationships and on their applications. For instance, if the world, if nature, is seen not only as an object of man's activity but as a creation, as a gift entrusted to man, then man should feel responsible for every intervention in this world. And more, a man will treat another man according to the absolute or relative value he is willing to ascribe to him.

In the same way the idea of a god or of nature will condition the encounter between science and religion.

From a diachronic and synchronic survey of the religious phenomenon, it seems to me that we can see in the previous relationships the center of fundamental values for a religious consciousness. This does not mean that a non-religious consciousness might not share parts of them.

Responsibility is certainly the religious ethical attitude which unifies the three levels of relationships.

1. Man is responsible for the fullness of life of another man, for the concrete possibility for him to be a man, and for the acknowledgment of his dignity, i.e., of his absolute and ultimate value and of his right not to be mistreated.

2. But man is also responsible for this world, for the life in it and for its survival. He has to find the equilibrium between the exigencies of progress and the limits of the earth's resources. If the world is for man, he has no right to destroy the bases of his own survival and of that of future generations.
3. Finally, the religious man finds the source of his own responsibility not in himself but in somebody else, in God, who has spoken to man in many ways and in many times throughout human history and in human cultures.

I consider that the ethical imperative of responsibility towards the life of mankind and of the earth is the most important demand religion can address to genetics, but also to many other sciences, either as pure research or when it becomes applied science.

Two Biblical Examples

A direct examination of two texts from the Judaeo-Christian religious tradition,
1. Gen 2, 15–17; 3, 1–11; and
2. Mo 9, 14–27,

can help us to deepen our appreciation of both the theme of the ethical imperative and the theme of human integrity. I have selected this tradition because I know it best and not from a desire to exclude other religious traditions, which are equally valuable as evidence about the religious character of man.

1. Exegetes today seem to accept that the story or myth of the first human sin is meant as a theological explanation of the presence of evil in the world; an explanation of why disorder, evil and suffering are present in a world which was created good by God. The cause of the chaos present in the cosmos is man himself, who wanted to change his relationship with the world. This world was entrusted to man – to Adam. The commandment not to eat of the tree of knowledge is a symbol of his non-lordship over the world. It is the limit which marks him, too, as a creature, as responsible to another Being. But Adam does not want to recognize that the command, the imperative which grounds ethics, comes from another. He wants to be, he himself, the origin of the ethical imperative, the one who decides what is good and what is evil. To eat the fruit of knowledge means, in this text, to take possession of the power to make decisions about the world. The world is no longer a gift to be cultivated and guarded but rather an object to be desired and possessed. Adam rejects the commandment which, precisely because it is a limit, is an active principle of order, and he constitutes himself as the principle of ethical judgement. The biblical writer traces the split between nature as created and nature as it de facto is back to the ethical autonomy of man. For the Bible the story of the human race is marked by this split and by attempts to repair it.

   The quality of the relationships between man and the world, between man and man, and between man and himself will depend on man's willingness to recognize a divine limit which defines him as an ethical being, accountable in the face of an imperative which he himself has not decided.

2. The gospels tell us of many miracles performed by Jesus Christ to heal diseases both physical and psychic, the so-called "demoniacs." In our context the story of the healing of the boy with epilepsy opens up opportune lines of reflection. Leaving aside the issue of the objective historical truth of the account, what interests us is the fact that Jesus Christ is consciously carrying out the messianic promises of the prophets which look forward to a mankind freed from diseases, including psychic diseases.

The cure of a sick man and therefore the scientific research that goes into curing him and into preventing diseases are not only not foreign to religion, they are an integral part of religious attitude. The aim of curing and/or of preventing sickness is to restore man to the wholeness of his physical and psychic being. To cure man is an ethical imperative, and a religious imperative as well, because the whole man, with all of his individual psycho-physical capacities, is man precisely as he was originally created.

Against the background of these two texts we can consider the whole field of genetics research and its clinical applications in a very positive way, and in particular the work of molecular genetics in psychiatry. Genetics is also a science of *how* a disorder is transmitted and is a science that intervenes in the nature of man as it is in order to bring it back to that state of nature as originally created which was intended by God the creator. But precisely because genetics intervenes in man, a living being in whom we recognize a dignity proper to his species in contrast to other living beings, at least in the western tradition, every genetic intervention takes on a special sensitivity. This fact should encourage prudence and a special respect in geneticists, but it cannot cast doubt upon the value of their work. This is especially clear if we consider the aim of the application of molecular genetics in psychiatry, which is to restore the individual or even future generations to their personal wholeness, a wholeness that must not be confused with a certain type of normality. But this topic could take us far afield.

Conclusion by Way of Postscript

Reviewing the presentations and the atmosphere at the meeting in Brno I believe that the following reflections are appropriate.

There is massive ignorance in mass media presentations of the clinical and ethical problems of molecular genetics research and its applications in the field of psychiatry.

Among the contributions of the participants in the meeting the present contribution stands out as different from the others both in its starting point and in the topic which it treats. This difference and the interest which it aroused confirm the value of the structure of this meeting decided on by the organizers, the WHO and Foundation IPSEN. This approach of interdisciplinary contact should be continued to overcome the risk of one-sided research on man and to overcome the danger of preventative or therapeutic procedures for mental illnesses which take no account of the integral structure of man. It seems that the presence of a "religions expert" will always be useful and profitable.

Nonetheless, even more useful and incisive than having a cultural or religious anthropologist at these meetings would be the presence of a moral theologian or of a moral philosopher. They could make their specific contributions starting from a framework of religious morality or of humanistic ethics, to the development of a deontological norm for the genetics researcher and for the clinical geneticist in psychiatry, a norm defining their ethical relations both towards the individual subject or patient and towards the society in which they work. It seems that more than one participant in the meeting voiced the demand for such a declaration of rights and responsibilities.

I would like, speaking personally, to express the wish that the religious attitude may always be of help to scientific research into this marvel which is the created world and man, and that the true scientific spirit will never exclude, due to prejudice, the religious dimension of man.

Jurisprudence in Genetics

R. Kimura

Introduction

It is quite clear that legislation is an instrument of social control that leads societies to change the function and system of their traditional ideas and behavior (Schubert 1975). Law is not identical to moral regulation and control, but legislation requires moral support if it is to be adjusted to the political, cultural, and economic framework of particular societies which actually produce new legislation (Miller 1979).

Due to the rapid development of biomedical science and biotechnology, as well as strong interest in their application to the public's interests in health, several achievements in the form of new legislation have been made by health experts and bureaucrats concerning public health issues (Brahms 1990).

In Japan, immediately after World War II, some of this legislation came into being followed rapid changes in society. In this paper I shall adopt the framework of past and present Japanese legislation, and consider prospective legislation in the form of case studies to clarify the basic jurisprudential issues that arise due to advances in modern genetics.

These topics are:

1. eugenic protection legislation and mental disabilities;
2. maternal-child health legislation and genetic screening; and
3. health information legislation and human genome analysis.

Eugenic Protection Legislation and Mental Disabilities

The Eugenic Protection Law (*Yusei Hogo Ho*) was promulgated on July 13, 1948, and remains in effect today. It took as its model the National Eugenic Law (*Kokumin Yusei Ho*) enacted in 1940. The enforcement of this National Eugenic Law rested on a clear policy of increasing Japan's population, in its quality and quantity, to serve as a base of state power under the influence of a wartime, military-oriented government bureau.

This legislation was influenced by the powerful ideology of the worldwide Eugenic Movement (Suzuki 1983). The aim of the legislation was to develop and promote a future Japanese population by employing eugenic screening processes to *prevent* an increase in the number of genetically inferior descendants,

Srám, Bulyzhenkov, Prilipko, Christen (Eds.)
Ethical Issues of Molecular Genetics in Psychiatry

including physically and mentally impaired descendants. The health of mothers was not much taken into account in this National Eugenic Law, although the words "Umeyo Fuyaseyo" (Be fruitful and multiply) became a national slogan for a state policy supported by the military regime in power during the war years, 1941–45 (Kimura 1984a, 1987).

After the war, due to a fundamental change in state policy – in transition from a military regime to a democratic political system based on a new constitution (Nov. 3, 1946) – and in the midst of social and economic confusion as well as an enormous population increase, the old National Eugenic Law (1940) was abolished and the new Eugenic Protection Law (1948) was enacted. Even though drastic political changes occurred in Japanese society, which led to the formation of a law relating to the health of the people, the term "eugenic" remained. Yet this new legislation had a completely opposing objective: to decrease the Japanese population by permitting abortion without prosecution only in cases of particular medical and social indications provided in article 14 of the Eugenic Protection Law. Indeed, even now, abortion remains illegal in Japan (Chapter 29, art. 212–216, Japanese Criminal Code, 1907).

Controlling the size of the Japanese population was also the policy of the Occupation Forces under the command of U.S. General Douglas MacArthur and his staff. The Eugenic Protection Law was proposed by a Japanese congressman, Dr. Ohta, who had been an advocate of family planning practices since the 1930s; however, there had been a good deal of pressure from the Natural Resource Section of the General Headquarters to enact this new legislation, and various comments concerning this proposed law appeared in official documents of the government (Kimura 1984a).

Thus the new Eugenic Protection Law was simply regarded as an abortion law rather than a eugenic law by medical professionals and the laypublic. One of the important articles in this law requires that there be no legal justification for abortion due to a genetically defective fetus. There has been a proposal to amend this article, but it has not yet been adopted.

The purpose of the Eugenic Protection Law is stated in Article 1:

> "The purposes of this law are to prevent the birth of inferior descendants from the eugenic point of view, and to protect the life and health of the mother as well."

The Eugenic Protection Law actually serves as a law for eugenic and maternal protection by applying the methods of "eugenic operation" (Article 2) and "artificial interruption of pregnancy" (Article 2-II). In this law the term "eugenic operation" refers to any surgical operation that makes a person unable to reproduce without removing the reproduction glands, as prescribed by Order. "Artificial interruption of pregnancy" refers to the artificial discharge of a fetus and its appendages from the body of the mother during the period when a fetus is unable to remain alive outside the body of the mother. (This particular time period ends around the 22nd or 23rd week of gestation.)

It is quite important to note that the original text of the Eugenic Protection Law included no provisions permitting economic and social reasons as justifications for an artificial interruption of pregnancy. In 1949 the law was amended to remove the rigid provisions concerning maternal protection, with the qualifying statement that: "if the mother's condition is seriously endangered by economic reasons ..." In 1952, another amendment was introduced to abolish any investigation by the District Eugenic Protection Commission and to give physicians the final authority to decide when to artificially interrupt a pregnancy, as well as to provide women a program for practical guidance in birth control and family planning.

The Eugenic Protection Law has three major elements:

1. articles relating to the process of eugenic operations
2. articles relating to the conditions for having artificial interruptions of pregnancy
3. articles relating to practical guidance for birth control methods and family planning.

According to the Eugenic Protection Law there are two forms of "eugenic operation." One is called the *discretionary* eugenic operation (Article 3), which the physician may perform at his discretion after he obtains the consent of the woman and the spouse. There are *five* clear indications for this operation:

1. the person in question, or the spouse, has hereditary psychopathia, hereditary physical disease, or hereditary malformation, or the spouse suffers from mental disease or mental disability;
2. a blood relative, within the fourth degree of kinship of the person in question or the spouse thereof, has hereditary mental disease, hereditary debility, hereditary psychopathia, hereditary physical disease, or hereditary deformity;
3. the person in question or the spouse thereof is suffering from leprosy, which is considered to be contagious for the descendants;
4. a mother whose life is endangered by conception or by delivery;
5. a mother who actually has several children whose health condition is feared to be seriously affected by any future delivery

The second method is called the *non-voluntary* eugenic operation. If, as the result of an examination, the physician discovers one of the diseases enumerated in Table 1 and recognizes that a eugenic operation is necessary for the sake of the *public interest* to prevent the inheritance of the disease, he is required to report this finding to the Prefectural Eugenic Protection Committee (PEPC).

In cases of eugenic operations and sterilization, there are fundamental and crucial elements which are closely related to the protection of the rights of the individual (Brakel and Rock 1971; Wexler 1980). This is why the Japanese Eugenic Protection Law includes several articles that provide for appeals to challenge the first decision taken by the PEPC, and the second decision taken by the Public Health Council (PHC), and finally open the way for the initiation of a lawsuit in civil court, as provided in article 9 (see Fig. 1).

Very few of these non-voluntary operations are being performed (see Table 2), and these are mainly requested by parents, guardians, or spouses. Such requests

Table 1

1. Hereditary Psychosis Schizophrenia Manic-depressive psychosis Epilepsy 2. Hereditary mental deficiency 3. Remarkable mental psychopathology Remarkable abnormal sexual desire Remarkable criminal inclination 4. Remarkable bodily illness Huntington's chorea progressiva Hereditary spinal ataxia Hereditary cerebellar ataxia Progressive muscular atrophy Dystrophia musculorum progressiva Myotonia Congenital musculorum atonia Congenital cartilaginous malgrowth	Leukosis Ichthyosis Multiple soft neurofibroma Sclerosis nodosum Edidermolysis bullosa hereditaria Congenital porphyrin urine Keratoma palmare et plantare hereditarium Atrophia nervi optici hereditarium Pigment degeneration of retina Achromatopsia Congenital nystagmus Blue sclera Hereditary dysacousia or deafness Hemophilia 5. Intense hereditary malformation Rupture of hand, rupture of foot Congenital defect of bone

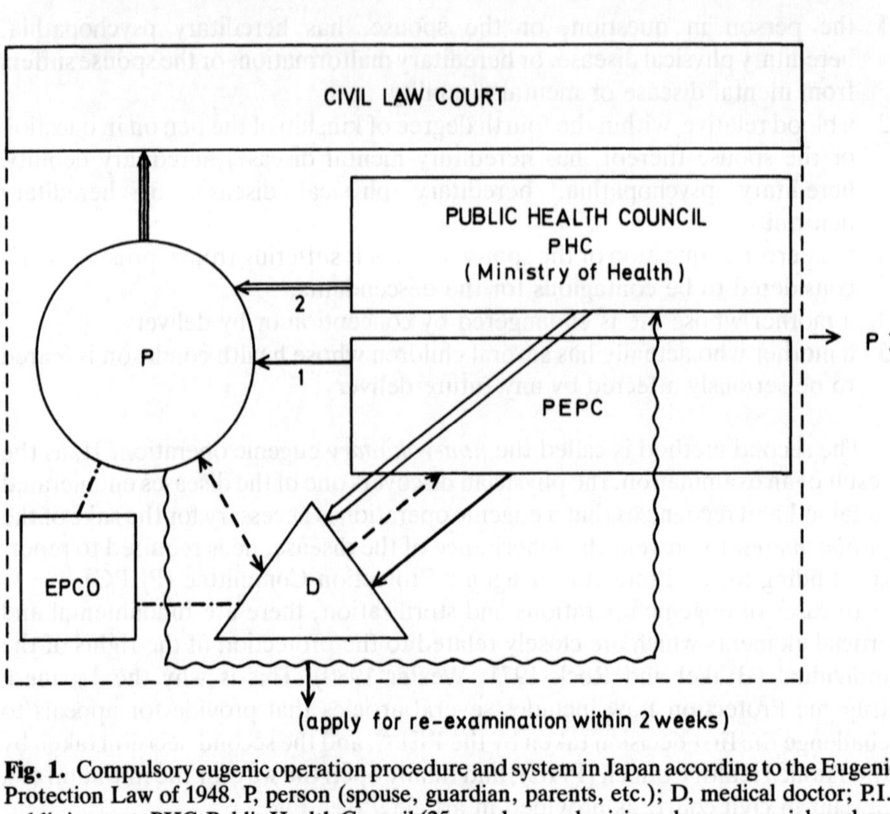

Fig. 1. Compulsory eugenic operation procedure and system in Japan according to the Eugenic Protection Law of 1948. P, person (spouse, guardian, parents, etc.); D, medical doctor; P.I., public interest; PHC, Public Health Council (25 members – physicians, lawyers, social workers, psychiatrists, etc.); PEPC, Prefectural Eugenic Protection Committee (10 members); EPCO, Eugenic Protection Consultation Office.

are made, as a matter of fact, due to eugenic considerations on behalf of a particular person. Non-voluntary operations are usually not performed if the person concerned brings the issue to civil court, because the majority of Japanese tend to think that courtroom resolution of these conflicts might not be socially appropriate (Kawashima 1963; Kimura 1988a).

One can raise a serious question concerning the fundamental ideological premises of the Japanese Eugenic Protection Law. One of the most problematic articles in this law concerns providing eugenic operation to patients who are in different categories than are indicated in the attached list (Table 1) of the Eugenic Protection Law, e.g., mental illness. Article 12 states:

> A physician may in regard to a person who is psychotic or mentally deficient (other than by hereditary causes mentioned in item 1 or item 2 of the Annexed List) obtain consent from the person who is responsible for protecting the patient under the provision of article 20 (In cases where the guardian, spouse, person exercising parental authority, or the person under obligation to sustain becomes person obligated to protect patient) of the Mental Hygiene Law (Law No. 123 of 1950) or article 21 (In the case where the mayor of the city or the chief of the town or village becomes the person obliged to protect the patient) apply for investigation concerning the reasonableness of performing eugenic operations to PEPC.

Article 13 provides that if an application to the PEPC is made, PEPC shall investigate whether or not this patient suffers from a psychosis or mental deficiency based on Article 12, and whether or not the performance of a eugenic operation is necessary to protect the patient. Thus the PEPC decides the reasonableness of performing the eugenic operation and informs the applicant and those who give consent (provided in Article 12).

Even though there are legal mechanisms to protect the rights of mentally ill patients, there have been quite a few cases concerning violations of these rights in various institutions for the mentally ill. Observations and documented reports, as well as recommendations by an International Commission of Jurists in Geneva together with a nationwide campaign for the reformation of this situation in Japanese mental hospitals, led to an open debate on these issues (International Commission of Jurists 1985). As a result, an amendment to the Mental Hygiene Law was passed in 1987. This shows the grave importance of international cooperation concerning changes in the traditional system, not only regarding mental diseases but also in the area of policy. Today, the people still need much more information and better education concerning mental and genetic diseases (Hirano 1987, Grostin 1987).

Maternal-Child Health Legislation and Genetic Screening

In Japan, there are several genetic screening programs for newborns. The Maternal-Child Health Law (*Boshi Hoken Ho*) of 1966 was enacted for the maintenance and promotion of the health of mothers, neonates, infants, and children.

Table 2. Number of sterilizations reported from 1955 to 1967, by sex, and stated reason[a]

			Voluntary sterilizations				"Non-voluntary" sterilizations	
			Hereditary diseases of					
Year	Sex	Total	the person concerned	the relative	Leprosy	Protection of maternal health	Hereditary diseases	Non-hereditary mental diseases
1955	M	1,528	34	52	14	871	534	23
	F	41,727	271	134	115	40,402	726	79
1956	M	1,774	36	70	17	1,158	482	11
	F	42,711	243	105	88	41,504	726	45
1957	M	955	9	11	3	698	222	12
	F	20,727	98	47	13	20,187	360	22
1958	M	1,641	27	22	9	1,174	394	15
	F	40,344	165	120	63	39,324	633	39
1959	M	1,205	23	8	8	821	335	10
	F	38,887	161	81	47	37,988	563	47
1960	M	1,130	37	20	7	853	203	10
	F	37,592	201	74	58	36,637	567	55
1961	M	1,049	21	12	13	724	270	9
	F	34,434	182	57	33	33,561	544	57
1962	M	964	18	10	1	717	197	21
	F	31,470	125	49	5	30,763	459	69

Table 2. Number of sterilizations reported from 1955 to 1967, by sex, and stated reason[a] (Continued)

		Voluntary sterilizations					"Non-voluntary" sterilizations	
			Hereditary diseases of					
Year	Sex	Total	the person concerned	the relative	Leprosy	Protection of maternal health	Hereditary diseases	Non-hereditary mental diseases
1963	M	832	10	7	0	629	166	20
	F	31,834	121	32	72	31,102	460	47
1964	M	708	5	10	1	547	133	12
	F	28,760	97	36	10	28,207	346	64
1965	M	697	11	5	0	533	127	21
	F	26,325	125	25	9	25,801	309	56
1966	M	535	10	0	2	427	86	10
	F	22,456	93	40	15	21,971	272	65
1967	M	553	5	10	2	456	70	10
	F	20,911	93	32	21	20,463	251	51
Total	M	13,571	246	237	77	9,608	3,219	184
	F	418,178	1,975	832	549	407,910	6,216	696
	M + F	431,749	2,221	1,069	626	417,518	9,435	880
	%	99.99	0.51	0.25	0.14	96.70	2.19	0.20

[a] Data are based on statistics from the Japanese Ministry of Health and Welfare

One of the unique practices based on this law (Article 16) is the issuance of a Maternal-Child Health Notebook to all of those who register at local offices or health centers governed by local authorities. Any woman who becomes pregnant informs the local office (Article 15) to receive various medical and health services, which she must then record in her Maternal-Child Health Notebook each time she receives these services (Kimura 1986).

There is no precise provision regarding a genetic screening program under the Ministry of Welfare and Health (MWH) or local government. However, a practical ordinance from the MWH gives administrative and legal justification for a genetic screening program which was initiated by health experts and bureaucrats at the level of central and local government (Ohkura and Kimura 1989).

In 1977, the mass-screening tests for inborn errors of metabolism were performed on only 29.2% of all newborns; in 1984 this percentage grew to 99.6% (Health and Welfare Statistics Association 1986). These screening programs for inborn errors of metabolism include PKU, maple syrup urine disease, homocystinuria, histidinemia, and galactosemia. The screening program for cretinism began in 1979, and for neuroblastoma in 1985.

These practical medical and health services are justifiable given the present legal framework. However, the detailed information provided for these screening practices, and the final endorsement by the laypublic, should be considered seriously in the health education process in the community, in local health centers, and in local schools. The initiative taken by the central and local governments on these issues of mass screening has delicate implications, relevant to an individual's health and the importance of protecting the privacy of genetic information. In this sense, even though there is a very positive side to the promotion of maternal-child health based on legislation, the practical requirements that all pregnant women file a report with the local authorities might cause some uneasiness for those who are seriously concerned about the real meaning of *privacy rights* (Kimura 1984b).

The positive reaction of Japanese pregnant women to this government ordinance reveals that the benefits of receiving genetic screening could also be recognized as a right of pregnant women to receive services from the government (Ohkura 1984).

Health Information Legislation and Human Genome Analysis

There is growing concern regarding future research on, and the application of, human genome analysis. The information acquired from human genome analysis could radically change the traditional notions of medical and health services, since some illness situations will be predictable beforehand. Genetic information about ourselves could change our lifestyles, behaviors, habits, etc. (Kimura 1990a).

Health information legislation designed to protect personal privacy has not yet been proposed in Japan. However, such legislation should be passed before scientific and technical "fixes" become available; otherwise, violations of privacy could occur too often. Schools, employers, insurance businesses, and future spouses might claim that it is necessary for them to obtain genetic information to protect the persons concerned. What criteria should be established to give or not give genetic information to these people and organizations?

The control and maintenance of genetic information in Japan has already begun, e.g., the registration of all pregnant women. There is no guarantee that personal privacy will be protected, even though there are some articles in the law that provide penalties for disclosure of particular information acquired by health professionals in the process of conducting their services (Kimura 1990b).

Additional integrated legislation concerning the private nature of health information (particularly as it relates to the human genome) is clearly required.

International guidelines are needed prior to initiating new legislation in various nations. The centralization of genetic information would give those in control enormous power over a population, and this could lead to rather serious consequences.

"The right to be different" in pluralistic, multi-cultural nations in the contemporary world should be a fundamental principle in jurisprudence that must be maintained as a basis for future legislation.

Concluding Remarks

Genetic manipulation for the enhancement of the body or abilities in a direct way should not be recommended and might be prohibited by legislation. However, genetic manipulation for the cure of disease and suffering would be justified bioethically and legally. Public debate on these issues of genetic manipulation would be extremely helpful, including contributions from various disciplines such as jurisprudence, bioethics, religion and genetics. Open communication between experts and the lay public should be a basic factor in making public policy and regulation relating to genetic health issues (National Institutes of Health 1990; Kimura 1988a).

References

Brahms D (1990) Human genetic information: the legal implications. In: Chadwick D, et al. (eds) Human genetic information: science, law and ethics. Wiley, New York, pp 111–118

Brakel SJ, Rock RS (1971) The mentally disabled and the law. University of Chicago Press, Chicago

Gostin L (1987) Human Rights in Mental Health: Japan. The Harvard University/World Health Organization International Colaboratoring Center for Health Legislation, Boston

Health and Welfare Statistics Association (1987) The trends of national health HWSA, Tokyo, p 104

Hirano R (1987) Psychiatry and law (in Japanese). Yuhikaku, Tokyo

International Commission of Jurists (1985) Human rights and mental patients in Japan. ICJ, Geneva, pp 80–86

Kawashima T (1963) Dispute resolution in contemporary Japan. In: von Mehren AT (eds) Law in Japan. Harvard University Press, Cambridge, pp 41–59
Kimura R (1984 a) The roots of family planning and its perspectives in Japan. Jpn J Nurs 48 (11): 1301–1304
Kimura R (1984 b) The meaning of gene therapy. Jpn J Nurs 48 (9): 1061–1064
Kimura R (1986) Caring for newborns. Hastings Center Rep 16 (4): 22–23
Kimura R (1987) Bioethics as a prescription for civic action: The Japanese interpretation. J Med Philos 12: 267–277
Kimura R (1988 a) Bioethics in the international community. In: Bernard J, et al. (eds) Human dignity and medicine. Elsevier, New York, pp 191–196
Kimura R (1988 b) Bioethical and socio-legal aspects of the elderly in Japan-with special reference to life-sustaining technologies. In: Institute of Comparative Law (ed) Law in East and West. Waseda University Press, Tokyo, pp 175–200
Kimura R (1990 a) Religious aspects of human genetic information. In: Chadwick D, et al. (eds) Human genetic information: science, law and ethics. Wiley, New York, pp 148–166
Kimura R (1990 b) The project for human genome analysis and bioethics, J Human Sciences 3 (1) Waseda University. Advanced Research Center for Human Sciences, Tokyo
Miller AS (1979) Social change and fundamental law. Greenwood, Westport
National Institutes of Health (1990) Governmental oversight and public participation. NIH, Washington, pp 6–7 (Gene therapy for human patients, information for the general public, part 2)
Ohkura K (1984) Clinical genetics (in Japanese). Nippon Ijishinpo Sha, Tokyo
Ohkura K, Kimura R (1989) Ethics and human genetics in Japan In: Wertz D, Fletcher JC (eds) Ethics and human genetics: cross-cultural perspective. Springer, Berlin Heidelberg New York, pp 294–316
Schubert G (1975) Human jurisprudence. University Press of Hawaii, Honolulu
Suzuki Z (1983) Eugenics in Japan (in Japanese). Sankyo Shuppan, Tokyo, pp 99–186
Wexler N (1980) "Will the circle be unborn?" Sterilizing the genetically impaired. In: Milunsky A, Annas G (eds) Genetics and the law II. Plenum, New York

Ethics, Molecular Genetics, and Psychiatric Disorders

M. Gill

Introduction

This conference was organised by the WHO to foster debate, generate ideas and hopefully develop consensus in the complex area of ethics relating to the application of molecular techniques to psychiatric disorders. The range of participants was broad; experts from the fields of genetics, molecular biology, psychiatric genetics, jurisprudence and ethics were present. Many were subspecialists; for example in single gene disorders, or in complex disorders such as heart disease. This chapter will outline the themes which were prominent, arising from stimulating discussion over the two-day conference.

Current Molecular Research

To begin, I will briefly review current knowledge and theories in psychiatric genetics, concerning the two major psychotic disorders, schizophrenia and manic-depression. Both disorders are common and tend to cluster in families. It is generally accepted that most of this familial clustering is due to shared genes rather than shared environment (McGuffin 1988; Mullan and Murray 1989). Inherited factors, however, account for only a proportion of the liability to develop the illness. Apparent Mendelian transmission, seen in some families (Book 1953), is the exception rather than the rule, and could equally be explained by a combination of chance and the effects of a number of genes, each of small effect (Edwards and Watt 1989). Nevertheless, it is also possible that the disorder in some of these families is caused by mutations at single gene loci, even if the bulk of the disorder is caused by polygenetic factors. For a period in the late 1980s it appeared that linkage analysis might provide rapid answers by demonstrating single gene effects and at the same time identifying the location of the gene (Mullan and Murray 1989). There is now considerable doubt that this will come about, at least in the near future. This pessimism has been occasioned by a general failure to replicate some positive findings (Hodgkinson et al. 1987; Kennedy et al. 1988) and by the reappraisal and withdrawal of others (Kelsoe et al. 1989).

The task of interpreting the evidence in favour of single gene effects was discussed by Baron, who remains confident that his linkage results on the

Srám, Bulyzhenkov, Prilipko, Christen (Eds.)
Ethical Issues of Molecular Genetics in Psychiatry

X-chromosome in manic-depression indicate the presence of a gene aetiologically involved in the disorder, at least in the pedigrees he has studied (Baron et al. 1987). He described ways in which both false positive and false negative results can arise and points out, as has been stated by others (Clerget-Darpoux 1989), that false negatives should be more likely to occur than false positives. If single gene effects do exist, then they are most likely to be clearly detected in large, multiply affected pedigrees. Suitable pedigrees – i.e., those with, as far as can be ascertained, a single source of the disorder, with large sibships, and with multiple affected generations – are rare in the extreme, particularly in Western societies. One role for the WHO that was suggested at the conference, and also previously by Murray and Gill (1990), would be to supervise a worldwide search for such pedigrees and to aid in diagnosis, collection of samples, and the efficient use of the resource in the laboratory. To avoid duplication of effort, WHO would have to coordinate with the European Science Foundation and North American collaborations.

Is it possible to discuss ethical issues relating to psychiatric molecular genetics when so little is known? Indeed it has been questioned whether it is ethical or not to commit so many resources to linkage studies of what are essentially complex genetic disorders (Pelosi and David 1989). In the light of results to date, it is certainly not possible to consider the use of linked polymorphic markers as predictive tests in any families with psychiatric disorders, in the same way as they are used in single gene disorders such as Huntington's disease. This is not to say that progress in molecular genetic research might not make possible some other form of predictive testing, such as the presence or absence of a susceptibility allele, to aid empirical data upon which contemporary genetic counselling in psychiatric disorders is based. Considering the lack of knowledge in psychiatric molecular genetics, if an examination of ethical issues is to be carried out, are there other disorders that may serve as models?

Ethics and Other Complex Partial Genetic Disorders

The genetics of psychiatric disorders are similar to other complex genetic disorders such as heart disease and diabetes. Dr. Berg, in this volume, points out that the identification of important susceptibility genes will make it possible to detect those who have a genetic predisposition to coronary heart disease. He suggests that individuals with such a predisposition might be motivated to undertake preventative efforts, such as dietary measures and abstinence from smoking. The identification of susceptibility alleles for psychiatric disorders would represent a major advance in our understanding, but to use such information for accurate genetic counselling would require that:

1. most susceptibility loci be identified;
2. their individual contribution to risk be accurately measured; and
3. protective alleles, if they exist, be identified.

Even with all this information, only the genetic component will have been measured. The considerable variation in risk estimates occasioned by environmental effects would also have to be taken into account. What would a psychiatrist say to a young person who is genetically at some greater risk for psychiatric disorder? In the absence of preventative measures, such knowledge is probably of little value and likely to lead to disadvantage for the individual concerned; it may even precipitate or exacerbate the disorder. All the negative uses of such genetic knowledge would in no way be balanced by the possibility of prevention. The information would, however, be available and therefore is open to being abused. The abuse by societies of what is taken to be accurate scientific knowledge has some precedence in psychiatric genetics. Eugenics, as practiced in the United States and Germany earlier this century, involved scientists as well as politicians. Scientists who seek to obtain a better understanding of the major psychoses by molecular genetic means need to understand the possible implications of their findings. It is also in their own interest that they are able to defend the ethical basis of their work, so that individuals and families continue to cooperate (Pelosi and David 1989). It would be wrong, on the other hand, to seek to discontinue research because of such historical misuse. Friedmann (this volume) points out that these lamentable parts of our genetic history were not entirely or even largely abuses engendered principally *by* Science, but rather they were abuses *of* Science by other sociopolitical institutions.

Lessons From Other Neuropsychiatric Disorders

Although based on a small number of autosomal dominant-like pedigrees, there is general agreement that a gene conferring susceptibility to Alzheimer's disease exists on chromosome 21 (St George-Hyslop et al. 1987). This agreement exists despite the finding that segregation analysis cannot distinguish between a multifactorial model and a mixed model consisting of a single gene plus a multifactorial background. Both of these models, however, explain the data better than any single gene model alone. Thus the genetics of Alzheimer's disease closely resembles that of the other major psychoses. McGuffin et al. (this volume) suggest that empirical data, based on data from family and twin studies, are all that can be used in counselling individuals as to their risk, or that of their offspring, of developing the disorder. Modular data (Murphy and Chase 1975) are incomplete and uncertain and cannot be used for the foreseeable future.

The experience gained in using DNA markers in predictive tests for Huntington's disease has already been a valuable learning experience for psychiatrists (Crauford 1989). It demonstrates some of the problems that can arise with a maturity onset disorder involving neurological and psychiatric symptoms. Huntington's disease is an autosomal dominant disorder in which predictive tests, in suitable pedigrees, can be between 95 and 99 % accurate for the presence of the disease gene. The fact that there are no known environmental effects is a considerable advantage from the counselling point of view over the major psychoses. Yet there are difficulties in using data from linked probes, which cannot be 100 % accurate. Thus individuals still cannot be given a yes or no answer. To someone who has witnessed the disease process in a parent, a small risk can be difficult to assess unemotionally. For Huntington's disease it has been possible to draw up generally agreed guidelines for most counselling eventualities. Even in this genetically simple situation there is room for differences in opinion in relation to counselling issues. For example, a couple, one of whom has a 50 % chance of carrying the disease gene, seek testing for their foetus. The tests indicate that the foetus has the same risk as its parent i.e., 50 %. The parents decide to allow the conception to go to term. Later, the parent develops the disease. It is now known that the child will in due course become ill and this has happened without that child's consent. At the meeting, opinion was divided as to whether such testing should be allowed. Other difficulties concern the conflict between the rights of the individual versus the rights of the family, and possession of DNA and DNA-based information. If a test for the mutation is developed, will an individual have a right to such a test, even if he states in advance that, if positive, he will commit suicide?

In the case of Huntington's disease, using linked markers, 60 % of family structures are informative for testing (Farrar et al. 1988). Given a fully informative family structure, opinion among couples was sharply divided as to whether or not they would abort a foetus with the Huntington's disease gene (Kessler et al. 1987). How many couples would abort a foetus if it had a 10 to 20 % chance of developing a serious psychiatric disorder? Wertz (this volume) gives some indication of the numbers who might, and of the sizable percentage who think that it should be legally possible.

The Wider Debate

Sooner or later any debate on ethical issues widens to include such issues as the ethics of abortion and the use of foetal material for research or, more recently, for implantation into the brains of individuals with Parkinson's disease. It seems to some an anomaly that abortion is available in many countries essentially on demand, yet abortion for a reason (i.e., the sex of the foetus) is usually illegal (Verma, this volume). Similarly, an aborted foetus can be thrown in the incinerator, but many object to using that foetus for scientific investigation. Dr. Berg (this volume) points out that objectors have a duty to formulate their thinking clearly before condemning any research use of aborted foetuses.

For those involved in ethics, molecular genetics is about to present a new ethical minefield, namely the application of genetic engineering to the treatment of human genetic disorders. Dr. Friedmann (this volume) outlines the rapid advances in genetic engineering techniques and in the methods used to deliver the engineered gene to its target. He considers that such work is ethically justified in attempting to design methods of treatment or prevention of genetic disorders. There appears to be a useful boundary in this debate, between somatic cell and germ cell engineering. The former comes under general guidelines for experimental and potentially hazardous treatments. For the present there is general agreement that the latter is to be avoided, on the basis that it is unethical to impose any risk, no matter how small, on future generations. Future generations are also the concern of Dr. Sram (this volume), who examines the possible toxic effects of environmental pollution on genes.

Ideals and Reality

Ethical debate on the use of DNA data in psychiatric genetics, and indeed on the wide range of human genetic disorders, is overshadowed when global issues are considered. Genetic counselling is not available, let alone DNA testing, for the vast majority of people on this planet who could benefit from them. That genetic testing should be available to all is a noble goal in principle but unobtainable in practice. Even in the United States and the UK, genetic counselling is utilised more utilised by the wealthy. In all societies, the guiding principle in the selection of individuals and families for counselling should be need.

Abuse of genetic knowledge takes many forms. For example, as mentioned above, in some parts of India illegal amniocenteses are used to determine the sex of a foetus, with subsequent abortion of females. The wide range in the responses among genetic counsellors who replied to Dr. Wertz's international survey provides an illustration of cultural differences in attitudes. To complicate matters further, therefore, scientists need to have some awareness of potential abuses of their discoveries in other cultures. Here, the WHO with its international experience, can help to foster such awareness, by promoting further meetings to keep debate on ethical issues topical and related to the most recent laboratory advances.

A proposed code of ethics was discussed, arising out of a previous WHO meeting held in Paris in May 1989, entitled "Genetic approaches in the prevention of mental disorders." The main points were as follows (with acknowledgement to Dr Wertz):

1. Equal access to genetic services regardless of ability to pay
2. Protection of patient choices in reproduction
3. Full disclosure of clinically relevant information. Psychologically sensitive information should be disclosed only in the context of full and supportive counselling
4. Counselling should be non-directive
5. Patients' privacy should be protected from employers, insurers and other institutions
6. Screening should be voluntary, with the possible exception of newborns when early treatment is available
7. Future generations should be protected from exposure to genetically toxic substances
8. Prenatal diagnosis should only be used for the diagnosis of defects in the foetus.

This proposed code of ethics does not relate *in particular* to people with psychiatric disorders. It provides a framework around which to build such a code, taking into account the special nature of psychiatric disorders. To have any meaning, such a code must not only outline ideal practice but must also be firmly rooted in the realities of a wider context, and thus should embrace such issues as the equitable provision of diagnostic and treatment services according to resources and the attitudes of many diverse societies to people with psychiatric disorders. Many present at the meeting expressed the opinion that, as patients with psychiatric disorders are especially at risk of abuse concerning genetic information, there should be special mention and possibly legal protection of their rights. The WHO could play a key role in providing a forum for further debate in this area. In Japan, the use of the law to ensure that as many children as possible have the benifit of early postnatal identification of treatable illnesses, causes concern to some (Kimura, this volume), who point out that it removes freedom of choice. Individuals should be able to opt out, for whatever reasons, from any screening program, without incurring financial or other penalty. With the enormous potential for use or abuse of DNA-based screening, any society must be free to decide if it really wants to know.

Psychiatric geneticists have been criticised in the recent past for rushing ahead in the laboratory without giving due thought to ethical issues. They should not also be criticised for failing to make progress in providing measures of an individual's genetic risk. Society often gives conflicting messages to its scientists.

Although not separated in reality, the distinction can and should be made between using molecular biological tools in attempting to understand the biological factors which influence behavior, leading to rational attempts at the development of treatments, and the use of genetic information for all aspects of genetic counselling, or "way station technology," as described by Wertz (1990). It is probably of benefit to all that there be steady progress in the laboratory, side by side with debate on ethical issues and carefully thought-out guidelines for clinical practice.

References

Baron M, Risch N, Hamburger R, Mandel B, Kushner S, Newman M, Drumer D, Belmaker RH (1987) Genetic linkage between X-chromosome markers and bipolar affective illness. Nature 326:289–292

Book JA (1953) A genetic and neuropsychiatric investigation of a Northern Swedish population. Acta Genet Stat Med 4:1–100

Clerget-Darpoux F (1990) Genetic epidemiology strategies in psychiatric diseases. In: Bulyzhenkov V, Christen Y, Prilipko L (eds) Genetic approaches in the prevention of mental disorders. Springer, Berlin Heidelberg New York, pp 12–23

Crauford D (1989) Progress and problems in Huntington's disease. Int Rev Psychiatry 1:249–258

Edwards JH, Watt DC (1989) Caution in locating the gene(s) for affective disorder. Psychol Med 19:273–275

Farrar LA, Myers RH, Cupples LA, Conneally PM (1988) Considerations in using linkage analysis as a presymptomatic test for Huntington's disease. J Med Genet 25:577–588

Hodgkinson S, Sherrington R, Gurling H, Marchbanks R, Reeders S, Mallet J, McInnis M, Petursson H, Brynjolfsson J (1987) Molecular genetic evidence for heterogeneity in manic depression. Nature 325:805–806

Kelsoe JR, Ginns EI, Egeland JA, Goldstein AM, Bale SJ, Pauls DL, Long RT, Conte G, Gerhard DS, Houseman DE, Paul SM (1989) Reevaluation of the linkage relationship between chromosome 11p loci and the gene for bipolar affective disorder in the old order Amish. Nature 342:238–242

Kennedy JL, Giuffra LA, Moises HW, Cavalli-Sforza LL, Pakstis AJ, Kidd JR, Castiglione CM, Sjogren B, Wetterberg L, Kidd KK (1988) Evidence against linkage of schizophrenia to markers on chromosome 5 in a northers Swedish pedigree. Nature 336:167–169

Kessler S, Field T, Worth L, Mosbarger H (1987) Attitudes of persons at risk for Huntington's disease toward predictive testing. Am J Med Genet 26:259–270
McGuffin P (1988) Major genes for major affective disorder? Br J Psychiatry 153:591–596
Mullan M, Murray RM (1989) The impact of molecular genetics on our understanding of the psychoses. Br J Psychiatry 154:591–595
Murphy EA, Chase GA (1975) Principles of genetic counselling. Year Book Medical Publishers, Chicago
Murray RM, Gill M (1990) The potential and the pitfalls of molecular genetics for psychiatry: an overview. In: Bulyzhenkov V, Christen Y, Prilipko L (eds) Genetic approaches in the prevention of mental disorders. Springer, Berlin Heidelberg New York, pp 107–115
Pelosi AJ, David AS (1989) Ethical implications of the new genetics for psychiatry. Int Rev Psychiatry 1:315–320
St George-Hyslop PH, Tanzi RE, Polinsky RJ, Haines JL, Nee L, Watkins PC, Meyers RH, Feldman RG, Pollen D, Drachman D, Growdon J, Bruni A, Foncin J-F, Salmon G, Fromhelt P, Amaducci L, Sorbi S, Placentini S, Stewart GD, Hobbs W, Conneally PM, Gusella JF (1987) The genetic defect causing familial Alzheimer's disease maps on chromosome 21. Science 235:885–890
Wertz DG (1990) Ethical issues in the application of knowledge from molecular genetics to mental disorders. In: Bulyzhenkov V, Christen Y, Prilipko L (eds) Genetic approaches in the prevention of mental disorders. Springer, Berlin Heidelberg New York, pp 92–106

Subject Index

MIX
Papier aus verantwortungsvollen Quellen
Paper from responsible sources
FSC® C105338

If you have any concerns about our products,
you can contact us on
ProductSafety@springernature.com

In case Publisher is established outside the EU,
the EU authorized representative is:
Springer Nature Customer Service Center GmbH
Europaplatz 3, 69115 Heidelberg, Germany

Printed by Libri Plureos GmbH
in Hamburg, Germany